Author:
Tricia Ball, M.S. Ed.,
GATE/Mentor Teacher

Illustrator:
Kathy Bruce

Editors:
Evan D. Forbes, M.S. Ed.
Walter Kelly, M.A.

Senior Editor:
Sharon Coan, M.S. Ed.

Art Direction:
Elayne Roberts

Product Manager:
Phil Garcia

Imaging:
Alfred Lau

Research:
Bobbie Johnson

Photo Cover Credits:
Images © PhotoDisc, Inc., 1994

Publishers:
Rachelle Cracchiolo, M.S. Ed.
Mary Dupuy Smith, M.S. Ed.

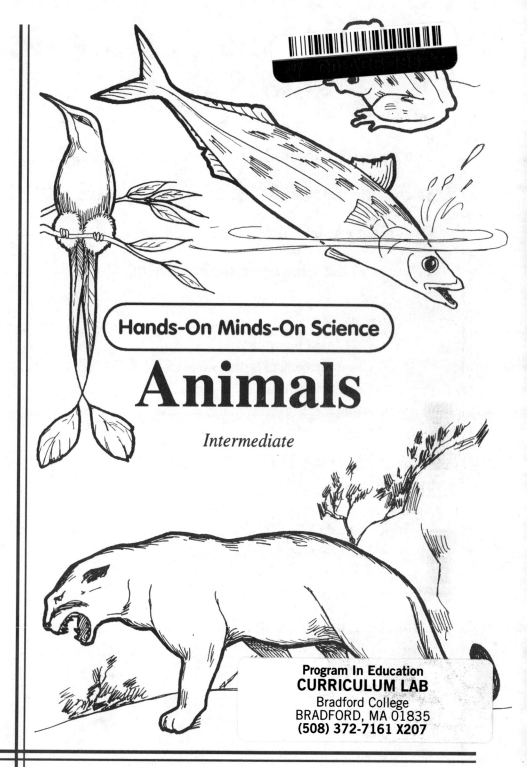

Hands-On Minds-On Science

Animals

Intermediate

Teacher Created Materials

Teacher Created Materials, Inc.
P.O. Box 1040
Huntington Beach, CA 92647
©*1994 Teacher Created Materials, Inc.*
Made in U.S.A.

ISBN-1-55734-630-5

Table of Contents

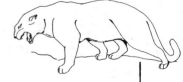

Table of Contents *(cont.)*

Introduction

Cur lab
QL
58
.B34
A64
1994

What Is Science?

What is science to children? Is it something that they know is part of their world? Is it a textbook in the classroom? Is it a tadpole changing into a frog? A sprouting seed, a rainy day, a boiling pot, a turning wheel, a pretty rock, or a moonlit sky? Is science fun and filled with wonder and meaning? What is science to children?

Science offers you and your eager students opportunities to explore the world around you and to make connections between the things you experience. The world becomes your classroom, and you, the teacher, a guide.

Science can, and should, fill children with wonder. It should cause them to be filled with questions and the desire to discover the answers to their questions. And, once they have discovered answers, they should be actively seeking new questions to answer.

The books in this series give you and the students in your classroom the opportunity to learn from the whole of your experience—the sights, sounds, smells, tastes, and touches, as well as what you read, write about, and do. This whole-science approach allows you to experience and understand your world as you explore science concepts and skills together.

What Is an Animal?

Scientists group plants and animals by the special characteristics of their body systems. They divided the living world into two categories: the Kingdom of Plantae (Plants) and the Kingdom of Animalia (Animals). In each of the kingdoms, plants and animals are further divided by their special characteristics. Using this method of classification, scientists are able to keep the living world identified in an orderly manner. Once you understand how scientists classify living things, you can understand how and why animals are able to survive the variety of places in which they live. In this unit you will learn what living things are, how scientists classify animals, and how animals are suited to the places in which they live.

The Scientific Method

The "scientific method" is one of several creative and systematic processes for proving or disproving a given question, following an observation. When the "scientific method" is used in the classroom, a basic set of guiding principles and procedures is followed in order to answer a question. However, real world science is often not as rigid as the "scientific method" would have us believe.

This systematic method of problem solving will be described in the paragraphs that follow.

1 Make an OBSERVATION.

The teacher presents a situation, gives a demonstration, or reads background material that interests students and prompts them to ask questions. Or students can make observations and generate questions on their own as they study a topic.

Example: Show students an animal.

2 Select a QUESTION to investigate.

In order for students to select a question for a scientific investigation, they will have to consider the materials they have or can get, as well as the resources (books, magazines, people, etc.) actually available to them. You can help them make an inventory of their materials and resources, either individually or as a group.

Tell students that in order to successfully investigate the questions they have selected, they must be very clear about what they are asking. Discuss effective questions with your students. Depending upon their level, simplify the question or make it more specific.

Example: What are living things made of?

3 Make a PREDICTION (Hypothesis).

Explain to students that a hypothesis is a good guess about what the answer to a question will probably be. But they do not want to make just any arbitrary guess. Encourage students to predict what they think will happen and why.

In order to formulate a hypothesis, students may have to gather more information through research.

Have students practice making hypotheses with questions you give them. Tell them to pretend they have already done their research. You want them to write each hypothesis so it follows these rules:

1. It is to the point.
2. It tells what will happen, based on what the question asks.
3. It follows the subject/verb relationship of the question.

Example: I think living things are made up of millions and millions of cells.

The Scientific Method *(cont.)*

4 Develop a **PROCEDURE** to test the hypothesis.

The first thing students must do in developing a procedure (the test plan) is to determine the materials they will need.

They must state exactly what needs to be done in step-by-step order. If they do not place their directions in the right order, or if they leave out a step, it becomes difficult for someone else to follow their directions. A scientist never knows when other scientists will want to try the same experiment to see if they end up with the same results!

Example: Once students know that cells make up living organisms, they will identify the major parts of an animal cell.

5 Record the **RESULTS** of the investigation in written and picture form.

The results (data collected) of a scientific investigation are usually expressed two ways—in written form and in picture form. Both are summary statements. The written form reports the results with words. The picture form (often a chart or graph) reports the results so the information can be understood at a glance.

Example: The results of the investigation can be recorded on a data-capture sheet provided (page 14).

6 State a **CONCLUSION** that tells what the results of the investigation mean.

The conclusion is a statement which tells the outcome of the investigation. It is drawn after the student has studied the results of the experiment, and it interprets the results in relation to the stated hypothesis. A conclusion statement may read something like either of the following: "The results show that the hypothesis is supported," or "The results show that the hypothesis is not supported." Then restate the hypothesis if it was supported or revise it if it was not supported.

Example: The hypothesis that stated "living things are made up of millions and millions of cells" is supported (or not supported).

7 Record **QUESTIONS, OBSERVATIONS,** and **SUGGESTIONS** for future investigations.

Students should be encouraged to reflect on the investigations that they complete. These reflections, like those of professional scientists, may produce questions that will lead to further investigations.

Example: Are all the cells in living things the same?

Science-Process Skills

Even the youngest students blossom in their ability to make sense out of their world and succeed in scientific investigations when they learn and use the science-process skills. These are the tools that help children think and act like professional scientists.

The first five process skills on the list below are the ones that should be emphasized with young children, but all of the skills will be utilized by anyone who is involved in scientific study.

Observing

It is through the process of observation that all information is acquired. That makes this skill the most fundamental of all the process skills. Children have been making observations all their lives, but they need to be made aware of how they can use their senses and prior knowledge to gain as much information as possible from each experience. Teachers can develop this skill in children by asking questions and making statements that encourage precise observations.

Communicating

Humans have developed the ability to use language and symbols which allow them to communicate not only in the "here and now" but also over time and space as well. The accumulation of knowledge in science, as in other fields, is due to this process skill. Even young children should be able to understand the importance of researching others' communications about science and the importance of communicating their own findings in ways that are understandable and useful to others. The animal journal and the data-capture sheets used in this book are two ways to develop this skill.

Comparing

Once observation skills are heightened, students should begin to notice the relationships between things that they are observing. *Comparing* means noticing similarities and differences. By asking how things are alike and different or which is smaller or larger, teachers will encourage children to develop their comparison skills.

Ordering

Other relationships that students should be encouraged to observe are the linear patterns of seriation (order along a continuum: e.g., rough to smooth, large to small, bright to dim, few to many) and sequence (order along a time line or cycle). By ranking graphs, time lines, cyclical and sequence drawings, and by putting many objects in order by a variety of properties, students will grow in their abilities to make precise observations about the order of nature.

Categorizing

When students group or classify objects or events according to logical rationale, they are using the process skill of categorizing. Students begin to use this skill when they group by a single property such as color. As they develop this skill, they will be attending to multiple properties in order to make categorizations; the animal classification system, for example, is one system students can categorize.

Science-Process Skills *(cont.)*

Relating

Relating, which is one of the higher-level process skills, requires student scientists to notice how objects and phenomena interact with one another and the change caused by these interactions. An obvious example of this is the study of chemical reactions.

Inferring

Not all phenomena are directly observable, because they are out of humankind's reach in terms of time, scale, and space. Some scientific knowledge must be logically inferred based on the data that is available. Much of the work of paleontologists, astronomers, and those studying the structure of matter is done by inference.

Applying

Even very young, budding scientists should begin to understand that people have used scientific knowledge in practical ways to change and improve the way we live. It is at this application level that science becomes meaningful for many students.

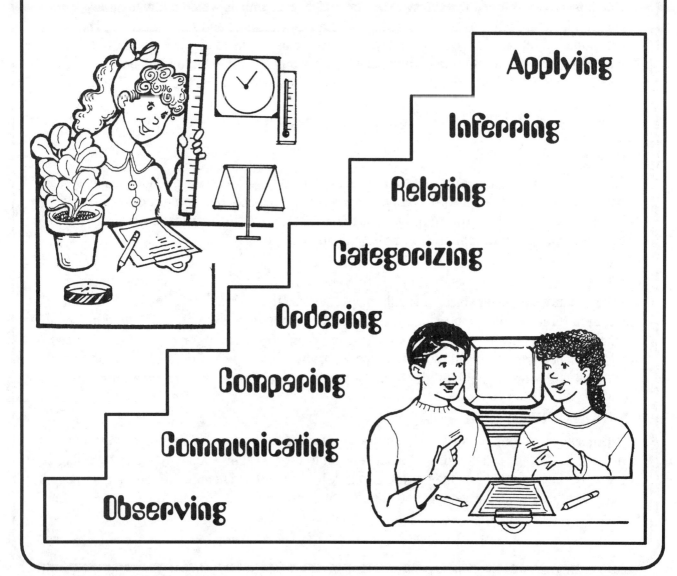

Organizing Your Unit

Designing a Science Lesson

In addition to the lessons presented in this unit, you will want to add lessons of your own, lessons that reflect the unique environment in which you live, as well as the interests of your students. When designing new lessons or revising old ones, try to include the following elements in your planning:

Question

Pose a question to your students that will guide them in the direction of the experience you wish to perform. Encourage all answers, but you want to lead the students towards the experiment you are going to be doing. Remember, there must be an observation before there can be a question. (Refer to The Scientific Method, pages 5-6.)

Setting the Stage

Prepare your students for the lesson. Brainstorm to find out what students already know. Have children review books to discover what is already known about the subject. Invite them to share what they have learned.

Materials Needed for Each Group or Individual

List the materials each group or individual will need for the investigation. Include a data-capture sheet when appropriate.

Procedure

Make sure students know the steps to take to complete the activity. Whenever possible, ask them to determine the procedure. Make use of assigned roles in group work. Create (or have your students create) a data-capture sheet. Ask yourself, "How will my students record and report what they have discovered? Will they tally, measure, draw, or make a checklist? Will they make a graph? Will they need to preserve specimens?" Let students record results orally, using a video or audio tape recorder. For written recording, encourage students to use a variety of paper supplies such as poster board or index cards. It is also important for students to keep a journal of their investigation activities. Journals can be made of lined and unlined paper. Students can design their own covers. The pages can be stapled or be put together with brads or spiral binding.

Extensions

Continue the success of the lesson. Consider which related skills or information you can tie into the lesson, like math, language arts skills, or something being learned in social studies. Make curriculum connections frequently and involve the students in making these connections. Extend the activity, whenever possible, to home investigations.

Closure

Encourage students to think about what they have learned and how the information connects to their own lives. Prepare animal journals using directions on page 84. Provide an ample supply of blank and lined pages for students to use as they complete the Closure activities. Allow time for students to record their thoughts and pictures in their journals.

Organizing Your Unit *(cont.)*

Structuring Student Groups for Scientific Investigations

Using cooperative learning strategies in conjunction with hands-on and discovery learning methods will benefit all the students taking part in the investigation.

Cooperative Learning Strategies

1. In cooperative learning, all group members need to work together to accomplish the task.
2. Cooperative learning groups should be heterogeneous.
3. Cooperative learning activities need to be designed so that each student contributes to the group and individual group members can be assessed on their performance.
4. Cooperative learning teams need to know the social as well as the academic objectives of a lesson.

Cooperative Learning Groups

Groups can be determined many ways for the scientific investigations in your class. Here is one way of forming groups that has proven to be successful in intermediate classrooms.

*   **The Expedition Leader**—scientist in charge of reading directions and setting up equipment.
*   **The Zoologist**—scientist in charge of carrying out directions (can be more than one student).
*   **The Stenographer**—scientist in charge of recording all of the information.
*   **The Transcriber**—scientist who translates notes and communicates findings.

If the groups remain the same for more than one investigation, require each group to vary the people chosen for each job. All group members should get a chance to try each job at least once.

Using Centers for Scientific Investigations

Set up stations for each investigation. To accommodate several groups at a time, stations may be duplicated for the same investigation. Each station should contain directions for the activity, all necessary materials (or a list of materials for investigators to gather), a list of words (a word bank) which students may need for writing and speaking about the experience, and any data-capture sheets or needed materials for recording and reporting data and findings.

Model and demonstrate each of the activities for the whole group. Have directions at each station. During the modeling session, have a student read the directions aloud while the teacher carries out the activity. When all students understand what they must do, let small groups conduct the investigations at the centers. You may wish to have a few groups working at the centers while others are occupied with other activities. In this case, you will want to set up a rotation schedule so all groups have a chance to work at the centers.

Assign each team to a station, and after they complete the task described, help them rotate in a clockwise order to the other stations. If some groups finish earlier than others, be prepared with another unit-related activity to keep students focused on main concepts. After all rotations have been made by all groups, come together as a class to discuss what was learned.

Just the Facts

THE CELL

Cells are the microscopic building blocks of living things. Cells are specialized structures. Each cell is a living organism that combines with other living cells to make an animal or plant. Each single cell is made up of even smaller parts. Each of these parts has its own function to keep the cell alive. A cell is the smallest part of a living thing. Most cells are made from the same structures. These structures are arranged in a specialized way so that all the parts work closely together. When two or more cells perform the same function, they form a tissue. The skin, hair, and bones of an animal are all tissues. Some tissues can have the same function, too. When tissues work together, they are called organs. The lungs, heart, and stomach are all organs. When the organs in the body of a living thing work together, they are called a system. Some of the systems found in living things are the respiratory, the circulatory, and the nervous systems. It is amazing how very complicated an animal is. It is more than just that cute, warm, fuzzy creature that we see in pictures.

PARTS OF THE CELL

Nucleus—The nucleus is a round body found in the center of the cell. The nucleus is the control center of all the cell's functions. It is very much like your brain. Within the nucleus there are specialized bodies that contain the genetic make-up of the animal. Characteristics such as the color of the hair or feathers and even the sex of the animal are controlled by the nucleus.

Nuclear Envelope—The nuclear envelope is a thin two-layer membrane that surrounds the nucleus. The function of the nuclear envelope is to protect the nucleus and allow nutrients from the cytoplasm to pass through it.

Cytoplasm—The cytoplasm is usually the name given to all the area outside the nucleus. The cytoplasm contains numerous structures that carry out the cell's basic functions. Some of these functions are the breaking down of food, converting proteins and carbohydrates, and storing energy. The cytoplasm has a jelly-like appearance.

Vacuoles—Vacuoles are the small clear areas sometimes seen in the cytoplasm. The vacuoles are the actual storage cabinets for the cell's food.

Endoplasmic Reticulum—The endoplasmic reticulum is a series of sacs and tubes which provide the communication system for the materials passing between the nucleus and the cell membrane.

Just the Facts *(cont.)*

PARTS OF THE CELL *(cont.)*

Mitochondria—The mitochondria is the powerhouse of the cell. This is the main place for energy production within the cell. Ninety percent of the cell's energy is developed there.

Golgi Complex—The golgi complex acts as the cell's storage and secretion depot. All the cell's nutrients pass through the golgi complex waiting to be sent to different parts of the cell.

Plasma Membrane—The plasma membrane is a soft thin membrane or covering that controls the movement of materials in and out of the cell. Nothing can enter or leave the cell without first passing through the plasma membrane. You can see the plasma membrane of a hen when you break open an egg. It is the clear, thin film-like lining attached to the shell.

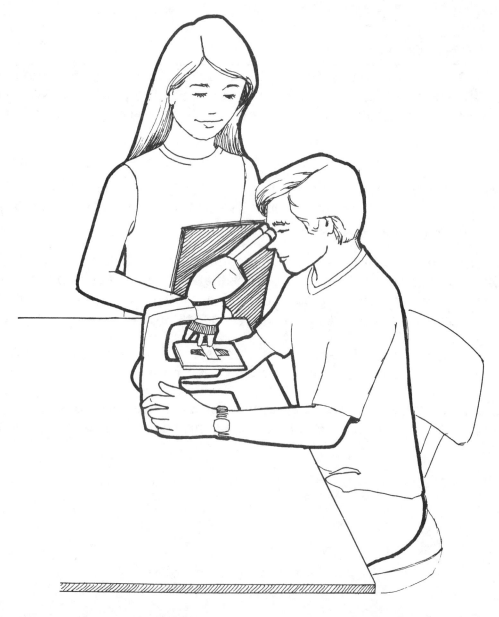

Up Close Look Inside

Question

What are living things made of?

Setting the Stage
- Ask students if they ever wonder what is below the surface of an animal's exterior.
- Have students share what they know about what is under the skin of animals. Create a list of the information they know. Keep this information on display throughout this unit.
- Discuss with students the outer coverings of animals. How do they look in the forest? Do they look the same in the jungle or the Arctic?
- Ask students if we can assume that animals having different characteristics on the outside will also look different on the inside. Have students discuss all possibilities.

Materials Needed for Each Individual
- variety of animal pictures (from around the world)
- crayons or colored markers
- data-capture sheet (page 14)

Procedure
1. Make an overhead of the animal cell on the data-capture sheet (page 14) and display for your class.
2. Distribute to students data-capture sheets with instructions (page 14).
3. Have students identify the following parts of the animal cell: *plasma membrane, nuclear envelope, cytoplasm, golgi complex, endoplasmic reticulum, mitochondria*, and *nucleus.*
4. Have students label and color the animal cell according to the key found on their data-capture sheets.

Extensions
- Have students find other pictures of cells. Explain that different parts of the animal are made from different types of cells.
- Have students compare the muscle cell, the skin cell, the nerve cell, and the cheek cell.

Closure

In their animal journals, have students draw and label each of the different types of cells found.

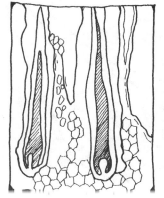 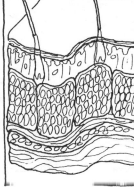  

Epidermis   **Muscle**   **Nerve**

Up Close Look Inside (cont.)

Identify the following parts of the animal cell. Label and color according to the following key.

Key

Plasma Membrane – Red

Cytoplasm – White

Endoplasmic Reticulum – Yellow

Nucleus – Orange

Nuclear Envelope – Blue

Gogli Complex – Purple

Mitochondria – Green

Close Encounters of the Microscopic Kind

Question

What is the smallest animal?

Setting the Stage

Ask students the following question. If you were a one-celled animal, what would you look like and where would you live? Accept any reasonable response.

Materials Needed for Each Group

- small jar of pond or creek water
- microscope
- microscope slide and slide cover
- eye dropper
- data-capture sheet (page 16), one per student

Procedure (*Student Instructions*)

1. Place a small drop of pond or creek water on your microscope slide.
2. Cover your sample and carefully place the slide under a microscope.
3. Observe what you see under the microscope and draw the organisms on your data-capture sheet.

Extensions

- Have students compare microscopic life found in a pond or creek to life found in a lake, ocean, or swamp.
- Purchase one-celled amoeba, paramecium, and euglena from a science supply house and have students study the three different organisms.

Closure

In their animal journals, have students compare and contrast their cell findings.

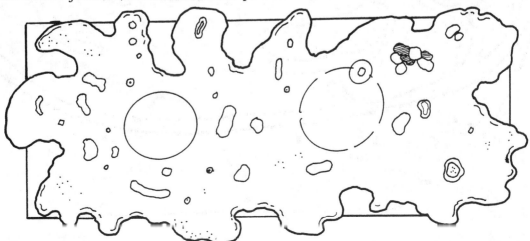

Close Encounters of the Microscopic Kind *(cont.)*

Draw a picture of what you see under your microscope.

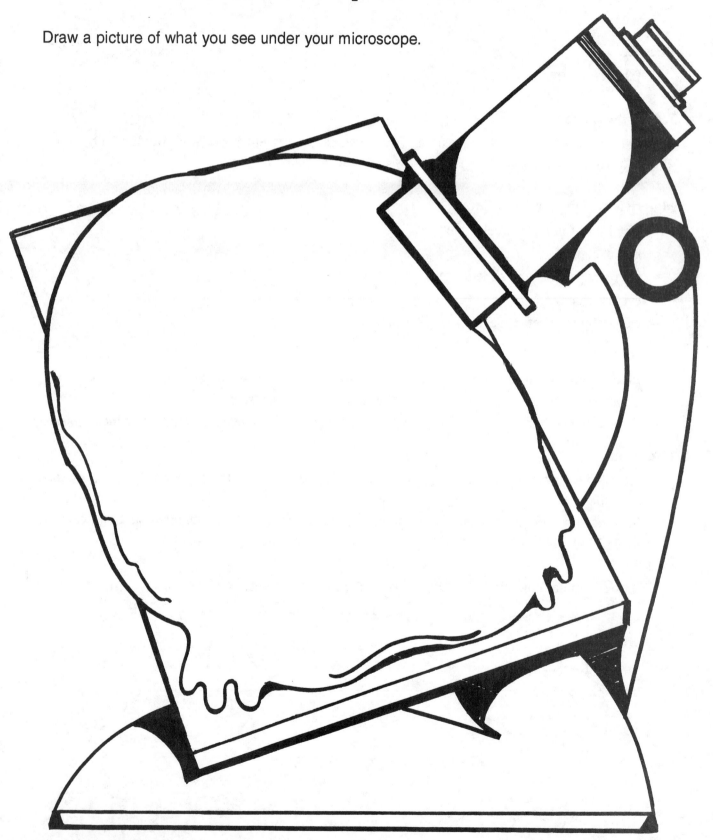

16

Cell Osmosis

Question

How does an animal cell get its food and nutrients?

Setting the Stage

- Discuss with students how people get the food and nutrients they need to survive.
- Discuss with students how plants get the food and nutrients they need to survive.

Materials Needed for Each Group

- an egg
- glass tubing
- tumbler, half-filled with water
- candle
- wax

   Note to the teacher: Have some extra eggs on hand in case of breakage.

Procedure *(Student Instructions)*

1. Chip the shell from one end of a fresh egg, being very careful not to break the white skin just under the shell. Make a small hole in the other end of the egg and put the glass tubing down into the egg. Do not break the yolk.

2. Light the candle and let hot wax drip around the tubing where it enters the egg in order to seal it tightly.

3. Put the egg in the glass tumbler so the exposed membrane is all underwater. When certain animal skins or membranes have solutions on both sides, an exchange takes place. The water will enter the apparently water-tight membrane and push the egg up inside the tubing.

4. This is how digested food gets into the blood stream, and how the root hairs of plants absorb water and plant food which must be in solution before osmosis takes place.

Extension

Have a cell biologist come to class and allow students to ask questions.

Closure

In their animal journals, have students draw a picture of their experience.

It's in the Bag!

Question

Can you make an animal cell in a bag?

Setting the Stage

Review with students the activity "Up Close Look Inside" (page 13).

Materials Needed for Each Individual

- small plastic zip-close bag
- 1/2 cup (125 g) of yellow gelatin
- 4-5 elbow macaroni
- 4-5 green peas
- round ball of clay
- 4-5 broken pieces of spaghetti about 1/2" (1.5 cm) long

Procedure *(Student Instructions)*

1. In each plastic bag add the above materials. It is that easy.

   The clay represents the *nucleus*.

   The macaroni represents *mitochondria*.

   The spaghetti represents the *endoplasmic reticulum*.

   The peas represent the *golgi complex*.

   The gelatin represents the *cytoplasm*.

2. You can add any other food to represent the other parts of a cell if you wish.

3. Paste a label to your "cell," listing and identifying all parts.

Extensions

- Have students repeat experience, this time making a human or plant cell.
- Have students create a display somewhere in the room of all their animal cells in a bag.

Closure

Compare your "cell" with those you saw under the microscope.

Cheek-to-Cheek

Question

What does a human cell look like?

Setting the Stage

Since we have been studying about the cell and its parts, ask students if they ever thought about the cells in their body and what they must look like. Do you think that they would be the same as those of nonhuman animals? Do you think they would be different? Why? Allow for all reasonable answers.

Materials Needed for Each Group

- microscope
- microscope slide and slide coverslip, one per student
- eye dropper
- toothpick, one per student
- water
- iodine
- data-capture sheet (page 20), one per student

Procedure *(Student Instructions)*

1. Using the eye dropper, put one drop of water on the slide.
2. Use the large end of the toothpick to gently rub the inside of your cheek. Do not use any toothpick except your own!
3. Roll the toothpick in the drop of water on the slide.
4. Add a drop of iodine to the water.
5. Gently place the coverslip on the drop of iodine and water.
6. Carefully place under the microscope.
7. Examine and record your findings on your data-capture sheet.

Extensions

- Have students view cells from the skin and muscle of different animals. They can use chicken, beef, or fish. Your local butcher or supermarket can help you obtain some samples.
- Have students do research on the various types of cells in the body. Ask them to identify the special function each different type of cell has.

Closure

In their animal journals, have students describe the functions of several different types of cells.

Cheek-to-Cheek (cont.)

Draw your cheek cell and label the parts: cytoplasm, cell membrane, nucleus

Answer the following questions.

What differences did you see when observing the cells under the microscope? High power? Low power?

_____

_____

Using your own words, describe what your cell looks like.

_____

_____

Compare your cheek cells to a piece of tissue paper and a leaf under the microscope. Draw the differences you see.

Cheek	**Tissue Paper**	**Leaf**

Just the Facts

Scientists have divided the world of living things into two major groups called kingdoms. They based their selection on the characteristics found in the cells. Animals cells differ from plant cells in that they have a plasma membrane and plant cells contain a rigid structure called a cell wall as well as a cell membrane. It should be noted that there are other kingdoms for those living things that do not meet the characteristics of plants or animals, but we will only discuss the animal kingdom here.

The animal kingdom or the Kingdom Animalia is probably the most recognized kingdom. Almost everyone can remember getting a stuffed bunny rabbit or having a favorite teddy bear. However there are some animals in the Kingdom Animalia you may not know about. Things such as sea cucumbers, slugs, sponges, and snails are also in this kingdom. The Kingdom Animalia contains all animals that lack cell walls.

The Kingdom Animalia is further sub-divided into smaller parts called *phyla.* The singular of phyla is *phylum.* Each of these groups is determined by the structure and function of the animals. Using the Phyla Wheel (page 24), label each phylum with its correct name, and then color the animals.

Phylum Protozoa—Consists of single celled organisms such as the amoeba.

Phylum Porifera—consists of those "pore-bearing" animals called sponges.

Phylum Coelenterata *(sill-enter-ah-tah)*—Consists largely of marine and aquatic animals ranging in size from a few inches (mm) to 2 yards (2 m) in length. These animals are usually colorful.

Phylum Platyhelminthes *(Platy-hel-min-theez)*—Contains flatworms and tapeworms, ranging in size from a few inches (mm) to more than 44 yards (40 m).

Phylum Aschelminthes *(ask-hell-min-theez)*—Contains round and unsegmented worms. The worms are usually parasites.

Phylum Annelida *(ah-nel-ih-dah)*—Contains aquatic worms and earthworms from 1 mm to 3 yards (3 m) in length.

Phylum Arthropoda—Contains more animals than all the other phyla combined. Arthropods are insects, spiders, crabs, lobsters, shrimp and all other animals with an external skeleton.

Phylum Mollusca—Contains soft-bodied animals like the ones found in the ocean or marine environment. Some mollusks are usually covered by a shell. They can live in fresh and salt water.

Phylum Echinodermata *(ee-ky-no-der-mah-tah)*—Contains large spiny marine animals. The sea star is one of the more common ones.

Phylum Chordata *(kor-dah-tah)*—Contains animals with backbones. Any animal having an internal skeleton is in this phylum.

Animal Collection

Question

Can you create an orderly system for keeping track of all the animals in the world?

Setting the Stage

- Discuss with your class the different characteristics that animals have.
- Have students list the many characteristics in the animal world on sheets of butcher paper or chart paper. Post around the room.
- Have students look in magazines and cut out and paste pictures of the many kinds of animals in the world.
- Have students brainstorm in groups to create a manageable way to organize and keep track of all the animals in the world.

Materials Needed for Each Group

- large piece of butcher paper or chart paper
- markers
- scissors
- glue or paste
- wide assortment of magazines and newspapers

Procedure

1. Have each group list all the animals they can think of in 15 minutes.
2. Have the same groups cut pictures out of magazines and newspapers to correspond to their lists.
3. Have each group report orally to the class about their findings.
4. After everyone has reported their findings, ask if they think there is a better way to organize the animal world.
5. Have students try grouping (classifying) their animals in several ways—e.g., fur-bearing, feathered, shelled, live-bearing, egg-bearing, animals with hooves, paws, tails, etc.

Extensions

- Have students investigate how scientists keep track of important information. Look in books for charts, graphs, tables, and lists that make the task of record keeping easier.
- Ask students to develop a way to keep track of a collection they might have at home. Examples: baseball cards, photographs, rocks, stamps, etc.
- Have some of the students share their collections.

Closure

In their animal journals, have students classify their list of animals in at least three ways.

Major Phyla

Question

Since there are so many different kinds of animals in the world, how do scientists categorize them?

Setting the Stage

- Have students evaluate the charts and posters they made in the animal collection activity.
- Ask students if they feel that anyone devised a really orderly manner to identify and record all the animals. Allow time for discussion.

Materials Needed for Each Individual

- copy of Major Phyla Wheel (page 24)
- scissors
- pieces of construction paper any color, 8.5" x 11" (23 cm x 28 cm)
- colored markers or crayons
- paper fastener

Procedure *(Student Instructions)*

1. Cut an 8" (20 cm) diameter circle out of construction paper.
2. Cut a 1" x 12" (2.5 cm x 30 cm) strip of paper out of construction paper.
3. Cut the phyla wheel out from page 24.
4. Measure the wedge in the wheel. Cut the same wedge out of your construction paper circle.
5. Place wheel over the construction paper circle and fasten with a paper fastener.
6. Place strip between the wheel and the construction paper. Print the names of each phylum.
7. Color the phyla wheel.
8. The numbers on the outside of the wheel represent the number of animals in each species.

Extensions

- Have students create a wheel for their own collections and share them with the class.
- Have students investigate the other kingdoms of plants and protists. Have them make phyla wheels for them.

Closure

Have students research the various characteristics of each phylum and why each animal was put in that category. Have them record this information in their animal journals.

Major Phyla (cont.)

Major Phyla Wheel

A. **Protozoa**

B. **Porifera**

C. **Coelenterata**

D. **Platyhelminthes**

E. **Aschelminthes**

F. **Annelida**

G. **Arthropoda**

H. **Mollusca**

I. **Echinodermata**

J. **Chordata**

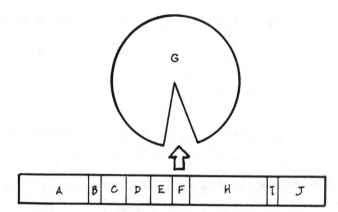

Similar and Different

Question

Can you group and compare animals using their similarities and differences?

Setting the Stage

- Have students get into groups and brainstorm a list of all the similarities and differences that can be used in identifying animals.
- Then on large chart paper make a master list for the class.

Materials Needed For Each Group

- wide variety of pictures of animals cut from magazines
- pencil and paper
- copy Dichotomous Key (page 26)

Procedure

1. Ask students in what ways animals are similar and in what ways they differ. Chart responses on the board.
2. Have students divide their animal pictures into two groups. Then divide each of those groups into two more groups, each time choosing one specific characteristic apparent in each group.
3. Tell students to work from general to specific. For example, from birds to types of beaks and from four-legged animals to number of toes.
4. Have students illustrate their findings.
5. Introduce the Dichotomous Key (page 26).
6. Have students plot their findings on the key.

Extensions

- Lead a game of "Twenty Questions." Then have students conduct the activity.
- Have students develop a dichotomous key for other groupings, such as characteristics in a biome.

Closure

- Pass out another dichotomous key and have students create a key for a grouping of their choice.
- In their animal journals, have students write a paragraph beginning, "I am unique! I am different from everyone else because . . . ."

Similar and Different *(cont.)*

The Dichotomous Key

Title

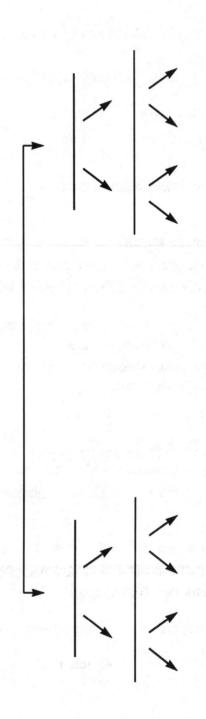

Microscopic Creatures

Question

Can you identify the microscopic creatures by their special characteristics?

Setting the Stage

- Show students several pictures of different plants and see if they can identify their special characteristics.
- Show students several pictures of different animals and see if they can identify their special characteristics.

Materials Needed for Each Group

- colored markers or crayons
- data-capture sheet (page 28), one per student

Procedure

1. Pass out data-capture sheets to students.
2. Review with students the special characteristics of the creatures pictured.
3. Allow students the opportunity to discuss the characteristics and match them to the creatures.
4. After matching, students can color their data-capture sheet.

Extension

Have each group imagine their own creature. Provide large size paper for them to draw and list their creature's unique characteristics.

Closure

Have the groups exchange pictures and try to provide a story about the creature. Where did it come from? What makes it unique?

Answer Key (page 28, "Microscopic Creatures")

1. amoeba	2. paramecium	3. tetrahymena	4. chilomonas	5. algae
6. euglena	7. hydra	8. stentor	9. planaria	10. vorticella
11. daphnia	12. rotifer	13. cyclops	14. brine shrimp	15. volvux

---

Microscopic Creatures *(cont.)*

Match the creatures with their descriptions and then color the page.

1. _____ I move slowly and look like a dab of jelly under a microscope. I have false feet or pods.

2. _____ I am a slipper-shaped animal with many hairy feet that help me swim.

3. _____ I am used in studies on nutrition because I can live in a liquid that is bacteria free.

4. _____ I am often considered to be half plant and half animal, and I puzzle scientists because of this feature.

5. _____ I am a plant having chlorophyll. Some people call me seaweed.

6. _____ I move by lashing my whip-like tail like a bullwhip.

7. _____ I am a freshwater relative of jellyfish and can grow new legs and arms.

8. _____ I am a small animal with a lot of lip.

9. _____ I am a flatworm and live in pond weeds and under the surfaces of dead leaves.

10. _____ I look like an upside-down bell with a tail. I was discovered over 300 years ago by Leeuwenhoek.

11. _____ I am a common water flea and have as many as seventeen instars or periods between molting.

12. _____ I have hair-like lobes that look like wheels on top of my head. I am often used to study life cycles and aging.

13. _____ I am an animal named after a lawless giant having one circular eye in the middle of the forehead.

14. _____ I am often called a "fairy shrimp" and have large dark eyes and numerous legs. I like it in darkness.

15. _____ I create waves as I just beat around my round self.

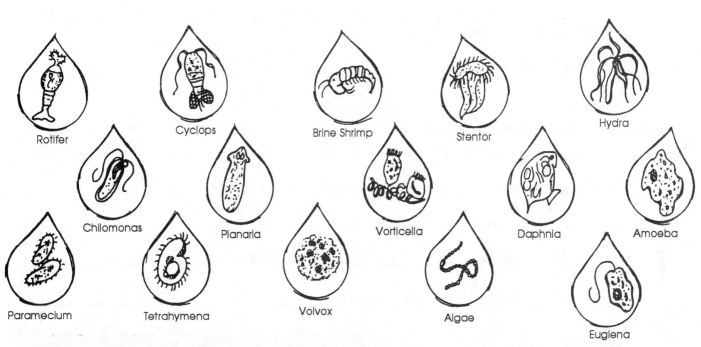

Rotifer   Cyclops   Brine Shrimp   Stentor   Hydra

Chilomonas   Planaria   Vorticella   Daphnia   Amoeba

Paramecium   Tetrahymena   Volvox   Algae   Euglena

Just the Facts

Far from the equator is the tundra. The tundra is cold and frozen most of the year. Temperatures can fall as low as -58° F (-50° C). The winter days are short and the nights long. There is very little sunlight in the tundra. Even in the summer, which lasts only six to eight weeks, the temperatures will reach only as high as 54° F (12° C). There is little precipitation (rainfall) anytime of the year. The soil found in the tundra is frozen most of the year. Scientists call this frozen soil the *permafrost*. Melting snow cannot seep into the ground, so little lakes and marshes appear during the summer months. The tundra extends across the upper part of Europe, Asia, North America, Russia, and Greenland.

The Arctic tundra is inhabited by many animals during the summer months. Grazing animals such as musk oxen and caribou live there. Polar bears live on ice floes and feed on fish. Animal life on the tundra thrives during the two summer months. Insects are everywhere. Flocks of birds and geese migrate to the marshes and the lakes for water. Small animals such as rabbits, wolves, mice, and snowshoe hares are found on the tundra. Resembling a mouse, a small creature called a lemming can be seen running on the tundra, too. Larger animals that migrate to the tundra during summer are the caribou and reindeer. It is during the summer that the animals feed on moss, lichens, grasses, and wild flowers. In the winter, things are much different. Many of the animals and birds migrate to warmer climates. When winter falls in the tundra, there are still a few animals that have adapted to the severe winters and live there all year long. The ptarmigan, ground squirrels, and the lemming are the only animals that stay. The lemmings survive by burying themselves in the ground before it freezes, and they hibernate all winter long. The only mammal which does not need to hibernate in the tundra is the musk ox. The oxen have a thick fur coat which insulates their bodies to protect them from the cold.

Mapping the Tundra

Color the tundra blue.

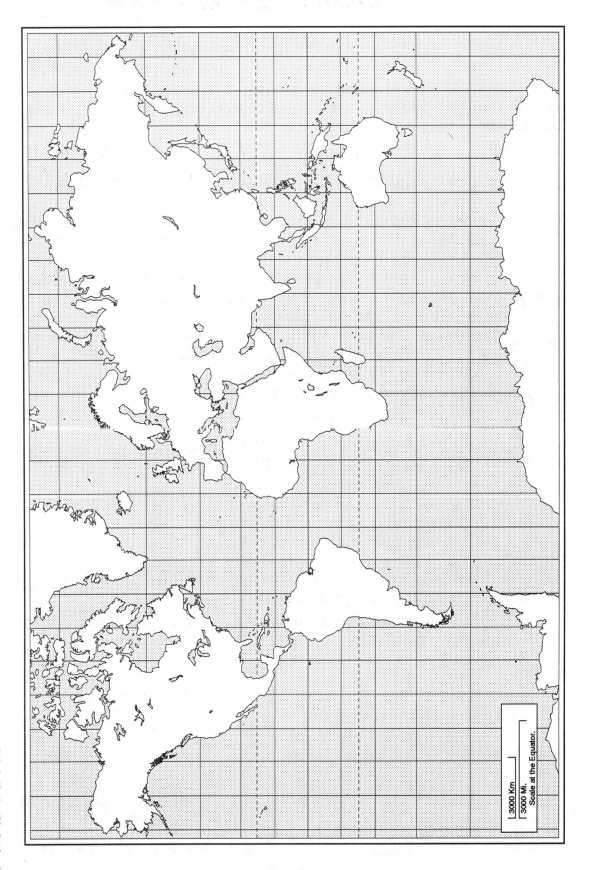

3000 Km
3000 Mi.
Scale at the Equator.

30

Solid as a Rock

Question

How do lakes and ponds form in the tundra?

Setting the Stage

Ask students how lakes and ponds can form if the tundra is frozen all year, even in the summer months. Allow for speculation and ask them to record their responses in their animal journals.

Materials Needed for Each Group

- pie pan
- soil (enough to fill one-half the pan)
- water
- tablespoon

Procedure *(Student Instructions)*

1. Fill each pie pan half-way with soil.
2. Place in a freezer overnight or until firm.
3. Pour 1-2 tbs (15-30 mL) of water over the frozen soil.
4. Observe carefully what happens on the frozen soil.
5. Record your findings in your animal journals. Compare them with what you thought would happen.
6. From your results, explain how the formation of the ponds and lakes in the tundra affect the animals that live there.

Extensions

- Have students research any other areas where this type of lake formation might take place and add this information to their animal journals.
- Have students research an animal that lives in the tundra for at least part of the year. Then report their research to the class.

Closure

In their animal journals, have students write a paragraph, imagining what it would be like to live in the tundra if they were an animal.

Sleeping—Tundra Style

Question

What is hibernation?

Setting the Stage

- Tell students in winter it is often too cold for animals to find food. As they have learned, the tundra is a frozen wasteland in the winter. Some of the animals survive these harsh winters by hibernating.

- Ask students what hibernation is. Hibernation is a long sleep, but the animals do not just go to sleep like us. They first prepare for hibernation by eating foods that will allow them to store fat in their bodies. Then they find a safe place to spend the winter. The little lemmings burrow into the ground. Finally they fall into a long, deep sleep. They spend the winter in hibernation until the days grow longer and the weather warms with the coming of spring.

- When animals hibernate, their body temperatures drop and their heartbeats slow.

Materials Needed for Each Group

- reference materials
- encyclopedias
- books on animals
- data-capture sheet (page 33), one per student

Procedure

1. Pass out the hibernation chart and allow students to complete it in groups.
2. This chart can also be used as authentic assessment since an understanding of hibernation is required.

Extensions

- Have students investigate what other animals hibernate and have them graph their findings in their animal journals.

- Have students write a dream an animal might have while in hibernation.

Closure

- Have students research the body temperature and pulse rate of normal sleep for animals on the hibernation chart. How do these data compare with the hibernation facts?

- In their animal journals, have students explain the differences between normal sleep and hibernation.

Sleeping—Tundra Style *(cont.)*

Body Changes During Hibernation

ANIMAL	AWAKE		IN HIBERNATION	
	Body Temperature	Heartbeats Per Minute	Body Temperature	Heartbeats Per Minute
HAMSTER				
BROWN BAT				
GROUND SQUIRREL				
WOODCHUCK				
BEAR				
LEMMING				

In your own words, explain what you found out. Record as many findings as you can.

_____

_____

_____

_____

_____

_____

_____

_____

_____

_____

Snug as a Bug

Question

How does a musk ox keep warm in the cold tundra winters?

Setting the Stage

- Remind students that musk oxen have a thick layer of fat inside their bodies.
- Discuss with students how we keep warm in the winter. (We stay indoors, we wear warm clothing, etc.
- Discuss with students the coarse hair that covers the oxen's bodies.
- Discuss with students insulation and how it keeps our homes warm.

Materials Needed for Each Group

- mixing bowl
- five plastic or metal containers
- insulating material—paper, wool, leather, styrofoam, aluminum foil, etc.
- thermometer
- gelatin
- large ice container or ice chest
- ice cubes
- data-capture sheet (page 35), one per student

Procedure *(Student Instructions)*

1. Cover all but one container with various types of insulating material.
2. Mix the gelatin and water. Fill each container, including the unwrapped one, with equal amounts of the gelatin mixture.
3. Place each of the gelatin mixtures into the ice container or ice chest. Record the beginning temperature.
4. Then record the temperature of the gelatin at set intervals. Choose 2, 3, or 5 minutes depending on your time frame. Make sure the temperature of the gelatin does not go above 41°F (5°C). If it does, add more ice, making sure that the ice does not go over the rim.
5. Observe which containers freeze first, second, etc. Compare your observations to the structure and body coverings of the musk oxen.
6. Construct a graph showing the relationship of the insulator to the rate at which it freezes.

Extension

Have students try other insulators. Beef fat obtained from a local meat market would be ideal. Also fur or animal skin is excellent for this experience.

Closure

- Have students explain how animals survive the winters.
- Have students compare the animals that stay in the tundra to the animals that migrate. How are they different and how are they similar? Ask them to write their observations in their animal journals.

Snug as a Bug *(cont.)*

Complete the chart and fill in the information needed.

Container No.	Beginning Temp.	Interval #1	Interval #2	Interval #3
Container 1				
Container 2				
Container 3				
Container 4				
Container 5				

1. Which container froze first, second, third, etc.? _____

   _____

   _____

2. Compare your observations of the containers, with that of the structure and body covering of musk oxen._____

   _____

   _____

   _____

3. Construct a graph showing the relationship of the insulator to the rate at which it freezes. Use the back of this page.

Just the Facts

The coniferous forest is the largest biome, spreading across North America, Northern Asia, and Northern Europe. Identify the area of the world that has coniferous forests on Mapping the Coniferous Forests, page 37. The coniferous forest is unique because of the types of trees growing there. The word *coniferous* means cone-bearing. In the coniferous forest you can find many of these cone-bearing plants such as the pine, hemlock, and spruce. In the coniferous forest the trees stay green all year, so they are called evergreens. There is a thick layer of pine needles which blanket the floor of the forest, providing homes for many animals. Just as the deciduous forest goes through the four seasons, so does the coniferous forest. The summers are shorter and drier. The winters are longer, colder, and wetter than those of the deciduous forest. The giant redwood trees found in Northern California are in the coniferous forest.

The Russians call the coniferous forest *taiga* or swamp forest because of the wet and muddy conditions found there from the melting snow during the spring and summer. There are many animals in the taiga. In the summer months, grazing animals live and feed in the meadows and clearings. These animals eat grasses and other plants. Moose and deer spend the day feeding on the shrubs and trees that grow there. Beavers live in ponds, rivers, and streams. There are squirrels and birds that live in the conifers that grow in the taiga. Larger animals like moose, wolves, and weasels find shelter in the forest. As in the tundra, you can expect to see a vast number of insects such as the butterfly living in the taiga.

The winter months in the taiga are long and cold. Many of the animals hibernate or migrate to other places. However there are some hardy animals that live in the taiga all year round.

Most of the animals that live in the coniferous forest have small bodies. Their bodies make it easier for them to move in the underbrush. You can find chipmunks, opossum, raccoons, skunks, and squirrels gathering food and running across the forest floor. These small animals are not the only animals in the coniferous forest. There are larger animals living there, too. Bear, deer, and moose share this unique biome. Along the shores of the rivers and ponds there are many animals that call the forest home. Beavers construct dams on the streams. Frogs live in the ponds and lakes, and turtles sun themselves along the banks of the rivers. Since the forest is damp, it makes a good home for many insects and the birds and animals that feed on them. One of the most unusual insects is the walking stick. It is found in North America and blends with the brown coloring of the bark found on the trees. The snowshoe hare is suited for life there because of its coloring and broad snowshoe-like feet. In the winter the hare is white to blend with the snow, but in the summer it turns brown to blend with the bark of the trees. It is not unusual for the musk oxen, elk, and moose to spend the winter in the taiga where food is plentiful.

Mapping the Coniferous Forests

Locate and color the areas of the world that have a coniferous forest. Use a green crayon for the coniferous forests to resemble the color of the evergreens.

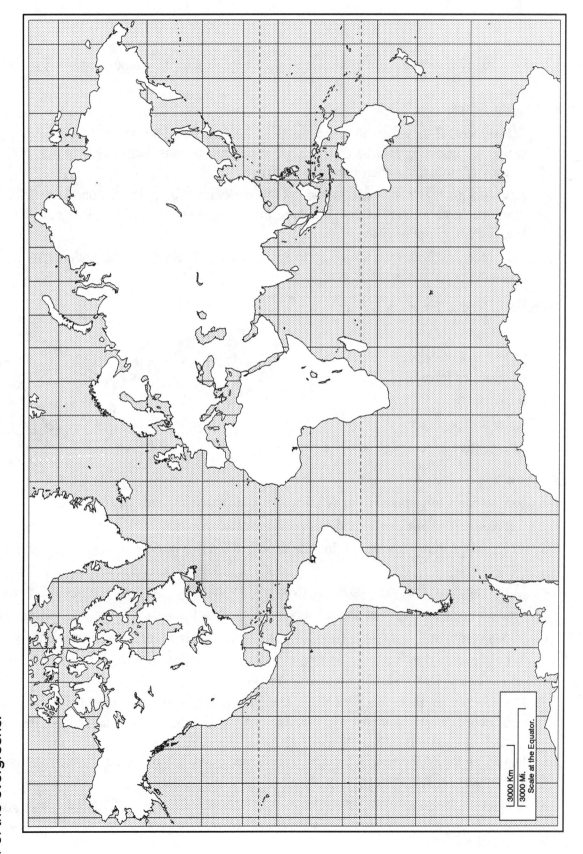

Hula Hoop Count

Question

How many insects do you think live in your backyard or the schoolyard?

Setting the Stage

- Tell students if they were to take a walk through a field on a warm spring or summer day, they would see insects everywhere. Crickets, grasshoppers, ants, butterflies, aphids, beetles—these and many more make their presence known.
- Ask students how many insects live in their backyard. Actually, in a one-acre field there may be more than one million insects. It is hard to imagine that there could be so many creatures buzzing and chirping away.
- Tell students, "Today we are going to do an activity that will help us understand how scientists count the populations of insects."

Materials Needed for Each Group

- hula hoop
- plastic jars or bags
- tweezers or forceps
- index cards
- paper and pencils
- data-capture sheet (page 40), one per student

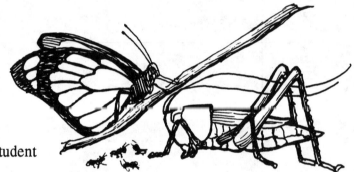

Procedure

1. Ask students to guess how many insects live in their backyards or schoolyards.
2. Record their guesses on your data-capture sheets.
3. Ask students how they would find the actual number of insects in an area. Accept any reasonable answer.
4. Inform students that there could be over 1,000 aphids, for example, on just one plant.
5. Find an open area, such as the schoolyard, a backyard, a vacant lot, or an open field.
6. Pick a common insect that is found in your area. In this example we will use the grasshopper.
7. Tell each team they are going to find out how many grasshoppers live in the field. A park is a good place to find insects if you live in the city.
8. Have each team toss out a hula hoop. Each hoop will be the boundary for that team. Each team must count the grasshoppers within the hula hoop boundary, They can collect the insects temporarily in their jars or bags to avoid counting the same ones twice. Have them check under leaves, on top of the soil, in the grass, on flowerheads, and in the air for their insects. Allow 15-20 minutes for this activity.
9. When time is up, have students release their insects and gather back together as a group. Have each group report their findings. The remainder of the activity can be done either on site or back in the classroom.

Hula Hoop Count *(cont.)*

Procedure *(cont.)*

10. Discuss with students the following questions:
    - Why did some plots have more insects than others? (Some areas might have had more food or fewer predators.)
    - Were there any problems encountered when trying to count your insects? (They keep hopping, flying away, or are too tiny to count.)
    - Can you think of a way to figure out how many insects there may be in the whole field? What else would you have to know? (They would need enough hula hoops to cover the field. They could add together all the grasshoppers from each hoop.)
    - Actually what scientists do is divide a field into square yards (m) with string. Suppose their field contains 15 square yards (m). They count the insects in each of the 15 squares and find the average. Then they multiply the average by the total number of square yards (m) in the field.
    - Remember this is only an estimated number.

Extensions
- Have each student make a bar, line, or circle graph to represent their insect count.
- Have a hula hoop count once a month for the school year.
- Construct a yearly graph showing the change in insect populations.

Closure

In their animal journals, have students draw pictures of the different types of insects found in their area. They can also construct models of the insects from papier mâché.

Hula Hoop Count *(cont.)*

Answer the following questions.

Answer before activity:

1. How many insects live in your backyard or schoolyard?

_____

_____

_____

2. How would you find the actual number of insects in an area?

_____

_____

_____

Answer after activity:

1. Why did some plots have more insects than others?

_____

_____

_____

2. What problems were encountered when trying to count your insects?

_____

_____

3. Can you think of a way to figure out how many insects there may be in the whole field?

_____

_____

_____

For the Birds

Question

What can people do to help the birds around their homes in the winter?

Setting the Stage

- Discuss with students the things birds need to have in order to survive.
- Ask students if those needs change during the winter.
- Ask students if a bird's food supply changes in winter. Is there anything they can do to help the birds out until the spring?

Materials Needed for Each Individual

- 1/2 gallon (2 L) milk container
- 1 yard (1 meter) of string
- scissors
- wild bird seed (optional)

Procedure *(Student Instructions)*

1. Cut out 2¹/₂" x 3" (6.25 cm x 7.5 cm) windows on all four sides of the container.
2. Cut 2.5" (6.25 cm) slits down the sides of each window. Fold to form a perch for the birds.
3. Poke six holes in the bottom of the container for a water drain.
4. Punch two holes at the top of the container for string.
5. Fill with bird seed, hang the feeder outside, and watch for the birds. It will not take them long to find your feeder.

Extensions

- Have students make numerous bird feeders and hang them around school. Provide seed for the various types of birds found in your area.
- Have a bird feeder contest. See who can construct the most original bird feeder. The prettiest. The most colorful.

Closure

In their animal journals, have students chart the numbers and kinds of birds served at their feeders for a week or a month and report the results of their findings to the class.

Coats and Boots

Question

How does camouflage protect animals in their environment?

Setting the Stage

Have students think about going to a wedding in a pair of cut-offs or going to the beach in their boots and winter coat. In both cases they would be out of place and easy to find in a crowd. Tell students, animals that do not want to be out of place and want to hide from their predators might want to camouflage their appearance. Camouflage is a technique used by animals to blend the shape, construction, or color of their bodies in with their surroundings for protection. Some animals might even resemble other animals for protection, while some animals such as the walking stick might resemble a twig or a part of a plant. This type of camouflage is called *mimicry*.

Materials Needed for Entire Class

- four boxes of toothpicks (wooden)
- two 6 x 6 yard (m) plots on the schoolyard field
- food coloring—green and brown

Teacher Preparation

- Soak one box of toothpicks in the brown and one in the green food coloring.
- Let them dry overnight.
- Cord off the 6 x 6 yard (m) plots the morning of the activity.

Procedure

1. Discuss with students the material in Setting the Stage.
2. Divide your class in two groups. Send one group out to scatter one box of colored toothpicks and one of plain toothpicks in the sectioned area.
3. When students have completed this, send the second group out to find as many toothpicks as they can.
4. Record the second group's findings.
5. Send the second group out to spread the two remaining boxes of colored and uncolored toothpicks. Repeat the process. Compare the two findings.
6. Have students answer the following questions: How are the results alike? How are they different? What factors might have affected their outcomes?

Extensions

- Have students repeat the experience, this time on a smaller scale with buttons of mixed colors.
- Have students find pictures hidden within pictures. Read *Where's Waldo?* (Martin Hanford, Little, 1990).

Closure

In their animal journals, have students respond to the following challenge: Suppose you wanted to hide a cut, unmounted diamond; using your knowledge of mimicry, where would you hide it? Explain why you chose this way of concealing your gem.

Do I Need a Map?

Question

Why do birds migrate?

Setting the Stage

- Tell students between two and five million birds migrate from rich feeding and breeding grounds in North America to warmer Central and South America before winter arrives. There are many kinds of migrating birds, including warblers, hummingbirds, and hawks. These birds can fly thousands of miles (km) in a few days. You might even live in an area where the birds migrate to. Some scientists feel that the birds use the stars to navigate. Others feel that the birds use the wind, the sunset, coastlines, or other landmarks to find their way from place to place. It is also possible that birds use the earth's magnetic field like built-in radar.

- Discuss with students the reasons that birds migrate in the winter. Lack of food, the cold, bad weather, scarce food, and even people are some of the reasons that birds migrate.

Materials Needed for Each Individual

- copy of "Why Do Birds Migrate?" (page 44)

Procedure

1. Distribute the "Why Do Birds Migrate?" activity sheet (page 44).
2. Have students pretend they are migratory birds such as the Canada goose, the swallow, or a warbler.
3. Arrange your students in groups of three and allow ample time for the game to be played.

Extensions

- Have students draw maps of the migratory routes the birds from the taiga travel in the winter.
- Have students investigate what other dangers face the birds during their migration.
- Have students read *Trumpet of the Swan*, a novel about the return of a flock of swans to their breeding grounds, by E.B. White, Harper C Child Bks, 1970.

Closure

In their animal journals, have students write a sample diary of what a flock of migratory birds might encounter on their flight south.

Do I Need a Map? *(cont.)*

Why Do Birds Migrate?

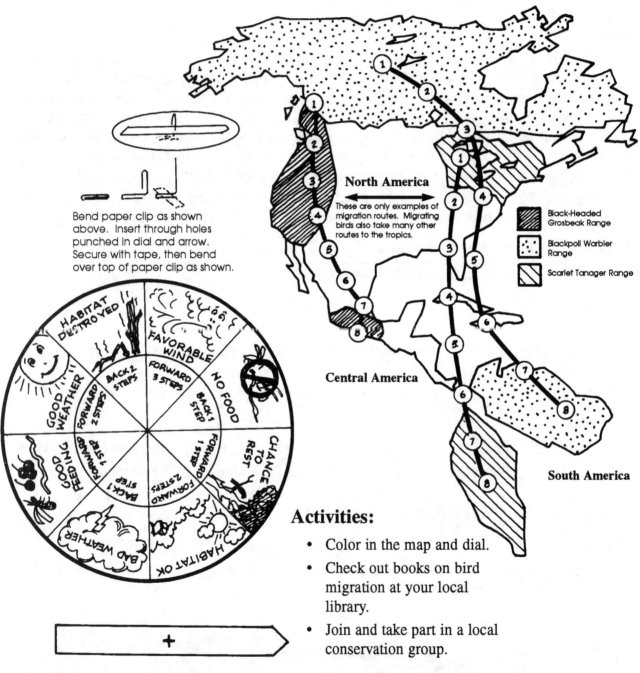

Bend paper clip as shown above. Insert through holes punched in dial and arrow. Secure with tape, then bend over top of paper clip as shown.

North America

These are only examples of migration routes. Migrating birds also take many other routes to the tropics.

Black-Headed Grosbeak Range

Blackpoll Warbler Range

Scarlet Tanager Range

Central America

South America

Activities:

• Color in the map and dial.

• Check out books on bird migration at your local library.

• Join and take part in a local conservation group.

In this game you can pretend you are a migrating bird. Place your player (bird) on step one. It takes eight steps to get from North America to the tropics. Taking turns, the three players spin the dial to see how many steps they may take. The first one to the tropics (step 8) wins the game! Use a dime and a penny or two buttons as your birds. To make dial cut out the arrow and dial from the skimmer or a xerox copy. Glue the paper to thin cardboard and cut out the pieces. Poke holes through the center of the dial and arrow, and using a paper clip, assemble as shown in the diagram above.

44

Just the Facts

The Deciduous Forest is located over much of the eastern half of the United States and western Europe. It is the forest of Rip Van Winkle and Robin Hood. In the deciduous forest, trees turn colors of red and gold in the fall and lose their leaves in the winter. Summers are generally hot. Spring and autumn are mild, and winters can be very cold. In the winters, the rain may fall as snow, but it usually melts as soon as it lands on the ground.

Since food for animals is abundant in the deciduous forest, there is also an abundance of animal life. Deer, raccoon, rabbits, and even bear roam the forest searching for food. There are also a great many insects in the forest. If you have ever hiked through the forest in the summer, you know that there are spiders and mosquitoes there too.

In the deciduous forest many lakes, ponds, and streams provide homes for the animals that live there. Frogs and salamanders bask on lily pads, and snakes hide under rocks and the underbrush. A familiar animal that can be heard hooting every evening in the deciduous forest is the owl. Owls are hunters, and investigating their pellets or droppings can tell you about their habits.

> In the southern hemisphere on the continent of Australia, you can find the forests where the echidna and the koala bear live.

Mapping the Deciduous Forests

Locate and color the areas of the world that have a deciduous forest. Use a red crayon for the deciduous forests to resemble the red leaves of fall.

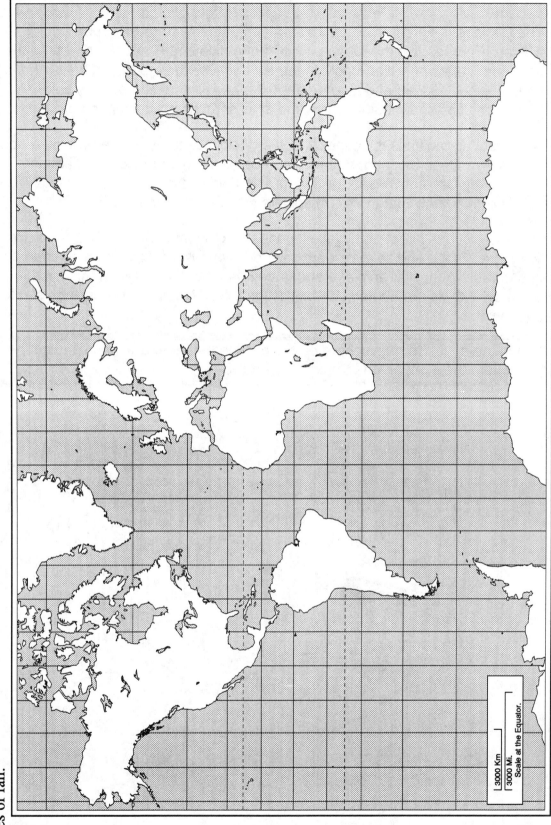

3000 Km
3000 Mi.
Scale at the Equator.

Animal Tracks

Question

How can we determine what animals live in our area?

Setting the Stage

- Draw for students a variety of animal tracks on the board and see if they can figure out what animals go with what tracks.
- Have students create an animal tracks guide.

Materials Needed for Each Individual

- cardboard or paper strips 24" (60 cm) long
- 8 oz (250 mL) milk container (top and bottom removed)
- mixture of plaster of Paris
- toothbrush
- copy of "Common Animal Tracks" (page 48)

Procedure *(Student Instructions)*

1. Go into your schoolyard, backyard, or field and locate animal tracks.
2. Put a collar of cardboard around the track that you find, or carefully place the milk container around the animal print.
3. Fill the container half-full with plaster of Paris and let dry to the touch.
4. When set, remove from the container, brush off, and you will have a negative cast.
5. Once you have your track, try to identify what animal made the track, by looking at the animal tracks on page 48.

Extensions

- Have students make a positive cast by greasing the cast with petroleum jelly and filling it with more plaster of Paris. This cast will show the animal's footprint exactly as it was found.
- Have students use the same procedure to make tracks of other items. They can try tire tracks, footprints, or even raindrops on sand.

Closure

In their animal journals, have students compare the animal tracks found to the types of animals found in different biomes. Can they determine the biome from the animal tracks?

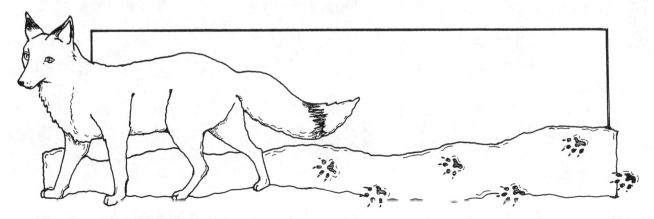

Animal Tracks *(cont.)*

Common Animal Tracks

Hind
Foot
Print

Squirrel

Hind
Foot
Print

Woodchuck

Oppossum

Otter

Muskrat

Skunk

Weasel

Fox

Racoon

Deer

Rabbit

Hup, Two, Three, Four, Six, or Eight

Question

Can you tell the difference between insects and spiders?

Setting the Stage

- Tell students spiders and insects are very closely related and easily confused. Elicit prior knowledge about spiders and insects. Show students pictures of spiders and insects. Have them make observations about the similarities and differences. Some general differences are below. They can be copied onto a chart for picture identification to follow.

Insects	Spiders
1. Usually have six legs.	1. Have eight legs.
2. Have three main body parts (head, thorax, and abdomen).	2. Have two main body parts (cephalothorax, with the head and thorax fused together) and the abdomen.
3. Found in water.	3. Usually live on land.
4. Have antennae.	4. Have no antennae.
5. Eat a variety of things, from plants to decayed material.	5. Usually are carnivorous and paralyze their prey with poison.
6. Most do not spin silk, and those that do usually spin it from glands in their mouths.	6. Most spin silk from their spinnerets on their abdomens.
7. Usually have two compound eyes and several simple eyes.	7. Usually have eight simple eyes and no compound eyes.
8. Usually have two pair of wings.	

- After the students have a good understanding of spiders and insects, do the picture postcard activity on page 50.

Hup, Two, Three, Four, Six, or Eight

(cont.)

Materials Needed for Each Individual

- slips of paper for each student in your class, half of them labeled SPIDER and half labeled INSECT
- one 4.5" x 5.5" (11.25 cm x 13.75 cm) piece of fairly stiff cardboard or a large index card

Procedure

1. Divide your class into two groups ("insects" and "spiders") by having each person pick a slip of paper.
2. Pair a "spider" with an "insect."
3. Send students to the library in small groups to find out about their animals. What do they look like? Where do they live? What do they eat? Students should be allowed to choose their own kind of insect or spider.
4. Using the cardboard, have students write letters to their partners. On the other side of the card they are to draw a picture of their insect or spider.

Extensions

- By putting themselves in the place of a spider or insect, the students will get a feel for what they need to live, grow, and survive. This activity can be used in each of the biomes in this unit. Students can send postcards to animals, birds, etc.
- Have students mail their postcards to others in the class.

Closure

In their animal journals, have students write three facts they have learned about their insects or spiders from their library research.

What Do Owls Eat?

Question

What can we learn about owls by studying their pellets?

Setting the Stage

- Have students define the term *raptor.*
- After students have a good understanding of raptors, have them list birds that they think fall under this category.
- Have a class discussion about owls.

Materials Needed for Each Group

- owl pellets (fumigated), commercially available
- tweezers or forceps
- magnifying glasses
- paper placemat to cover desks
- glue
- data-capture sheets (pages 52-53), one each per student

Procedure *(Student Instructions)*

1. Cover desk tops. Using the tweezers or forceps, carefully separate a pellet and spread its contents out on the desk.
2. Examine the skulls and individual bones found in the pellets. Compare them to the skeletons of small animals. Identify what the owl ate.
3. Count the number of skulls found in the pellet to determine how many small animals the owl ate in one meal.
4. On your data-capture sheet (page 52), draw the bones and skulls you found. Then try to draw a picture of the entire animal.
5. After counting and drawing what you found, glue the bones in the appropriate places on your data-capture sheet (page 53).

Extensions

- Having students use the estimate that an average owl will regurgitate two pellets a day. Have them explain how owls help to control the rodent population and what part they play in the predator-prey relationship.
- Have students draw a picture of an owl perched in a tree.

Closure

In their animal journals, have students write a paragraph responding to the following question: What can you do to provide a safe nesting area for a forest owl?

What Do Owls Eat? *(cont.)*

Draw a picture of the skulls you found.

Record the number found: _____.

Draw a picture of the bones you found.

Record the number of bones found: _____.

Reconstruct the skeleton of the animal found.

What animal might this be? _____

What Do Owls Eat? *(cont.)*

Glue the bones you find in the appropriate places.

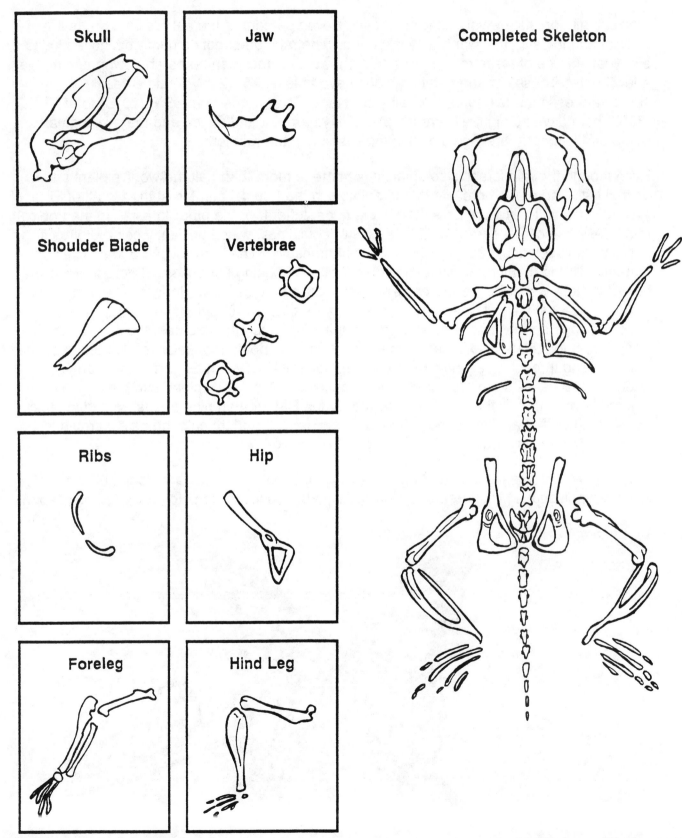

Skull	Jaw

Completed Skeleton

Shoulder Blade	Vertebrae

Ribs	Hip

Foreleg	Hind Leg

Just the Facts

The Tropical Rain Forest is located near the equator.

Tropical rain forests are very special places. Seven percent of the earth's land areas are tropical rain forests. It is warm all year round. The rain forest got its name because it rains everyday. Some of these rains can just be showers, or they can be harsh downpours that last a few hours. These hot and humid regions have at least 75" (2 m) of rain a year. Temperatures in the rain forest are fairly constant. They range from 68°F (20°C) to 85°F (30°C) both day and night all year long. The days are hot and moist, and the evenings are damp and warm. It rarely gets cool in the evening in the rain forest.

Even though the rain forest is so small, it is home to more than half the world's plant and animal life. There are more plants and animals in the tropical rain forest than in all of Europe! Animal life is quite different here than in the taiga, tundra, or the other forests. In the tropical rain forest, you can see trees as tall as 17 story buildings, watch brilliantly colored birds flying in the thick vegetation, and hear the chatter of monkeys. There are over 1,500 species of butterflies in the rain forest. Butterflies do not start out being butterflies. Before there is a butterfly, there is a fluffy, creepy, crawling insect called a caterpillar. In fact, a butterfly goes through a series of life stages called *metamorphosis.* Learn about metamorphosis in the Quick Change activity that follows. Like the animals of the other biomes, the animals in the tropical rain forest are also suited to their environment. Some monkeys and lemurs use their tails to swing through trees in search of food. Colorful birds such as parrots, toucans and budgies, or parakeets are protected by their brilliant colors. They blend into the brightly colored flowers that fill the rain forest all year long. Rather than hide themselves in the thick underbrush of the rain forest floor, many animals use their bright coloring as a warning to other animals to stay away.

Frogs in the rain forest come in many colors. Many are quite beautiful to look at, but their skins are poisonous. Hunters rub their arrows on the backs of frogs to create lethal weapons.

Just the Facts *(cont.)*

Another kind of creature found in the rain forest is the American click beetle. They measure only 1.5" (3.75 cm) in length, but you can hear their clicking sound throughout the world's rain forests. Click beetles are unusual in another way. They are a kind of firefly and have glowing abdomens. Learn just how these insects glow by completing "Bugs That Light the Night" (page 65).

In the rain forest there are animals that use their claws for grasping the trees, animals that use their tails for grabbing vines, animals with toes that stick to trees, and animals that fly and leap. Birds and butterflies are not the only animals in the rain forest that fly.

In the rain forests of North America, Asia, and Africa, there are flying squirrels and lizards. These animals have extra flaps of skin that enable them to catch the air and float on the air currents. The flying squirrels have skin flaps that extend from their front legs to their hind legs. The skin stretches out like a kite when they leap. The insect-eating lizards also have an extra flap of skin along the sides of their bodies that helps them glide through the air. It is important to remember that these animals are not fliers like the birds. They only use air currents and glide.

Mapping the Rain Forest

Since most of the rain forest is brightly colored, locate the rain forest on the map and color it yellow.

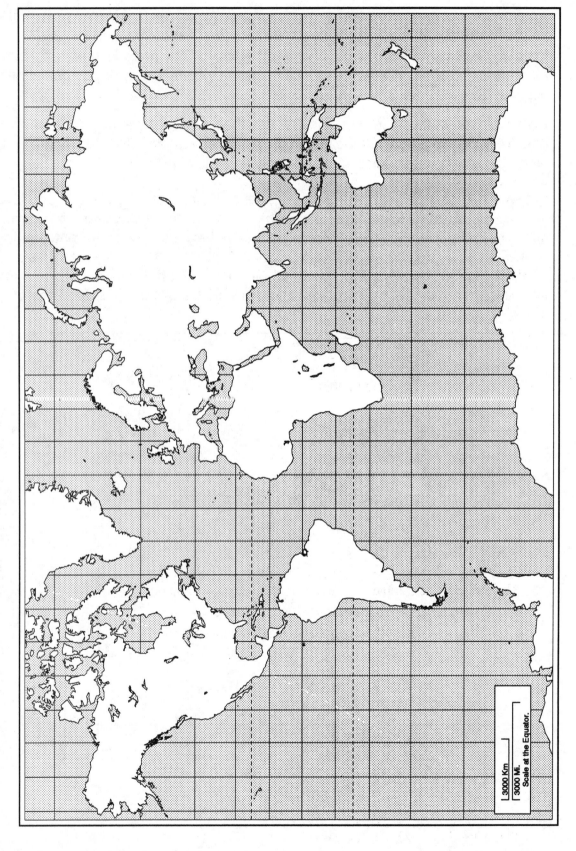

3000 Km
3000 Mi.
Scale at the Equator.

Butterflies and Moths

Question

Can we identify the butterflies or moths that live in our biome?

Setting the Stage

- Tell students the rain forest is alive with beautiful insects. They come in all colors and shapes. They might see a variety of these "lepidoptera" in their own backyards. Ask them if they know what these creatures are? They are the butterfly and the moth.

- Explain to students that the tropics are located near the equator and that it is difficult to travel there to experience the beautiful butterflies and moths first hand. In this activity we will become familiar with some common butterflies and moths and keep caterpillars in captivity. The above scenario will pique students' interest for the "Butterflies and Moths" activity.

Materials Needed for Each Individual

- a copy of "Butterfly and Moth Identification" (pages 58-61)
- scissors
- tagboard or cardboard for mounting
- colored markers or crayons
- books on butterflies and moths
- paper fasteners
- one sheet of construction paper

Procedure

1. Make a copy of "Butterfly and Moth Identification" for each student.
2. Have students cut out the pictures of the butterflies and moths and mount them on the cardboard.
3. Have students locate the butterfly and the moth in a book or encyclopedia and color them the correct colors.
4. Have students make a facts list on the rear of the card.
5. Have students cut a piece of construction paper to fit their cards and design a cover for their collection or use the one provided.
6. Have students fasten their cards and cover together to make a booklet.

Extensions

- Have students use the drawings of butterflies and moths on pages 58-60 to identify the ones that live in their biome.

- Have students identify the differences between a moth and a butterfly. Can you tell just by looking? What other facts do you need to know?

Closure

- In their animal journals, have students record and respond to the following questions: What impact does the butterfly have on our environment? What would be different if there were no butterflies?

- Have them read some books on butterflies and moths.

Butterflies and Moths (cont.)

Butterfly Identification

Monarch Butterfly

Common Sulphur Butterfly

Buckeye Butterfly

Cabbage White Butterfly

Painted Lady Butterfly

Tiger Swallowtail Butterfly

Butterflies and Moths *(cont.)*

Butterfly Identification *(cont.)*

Monarch Butterfly

Wingspan: 3.5 to 4" (8.75 to 10 cm)
Range: Includes the entire United States

This butterfly is active during the day. In the fall it travels south or migrates for the winter. In the spring, the monarch mates and then begins its return trip northward.

Common Sulphur Butterfly

Wingspan: 1.5 to 2" (3.75 to 5 cm)
Range: Includes all of United States except for many parts of Florida

This butterfly is active during the day. The common sulphur can be found in most open spaces. It likes farmland because of its eating habits.

Cabbage White Butterfly

Wingspan: 1.25 to 2" (3 to 5 cm)
Range: Includes the entire United States and Hawaii

This butterfly is active during the day. It likes fields and gardens.

Buckeye Butterfly

Wingspan: 2 to 2.33" (5 to 6 cm)
Range: Includes the entire United States

This butterfly is an adult during the winter months.

Painted Lady Butterfly

Wingspan: 1.75 to 2.5" (4 to 6.25 cm)
Range: Includes the entire United States

This butterfly is active during the day. It likes open places. The painted lady also migrates or travels south for the winter.

Tiger Swallowtail Butterfly

Wingspan: 3.25 to 5.5" (8 to 13.75 cm)
Range: Includes central and eastern United States

This butterfly is active during the day. It lives in many places, including woodland clearings, parks, and gardens. It can also be found by roadsides and rivers.

Butterflies and Moths (cont.)

Moth Identification

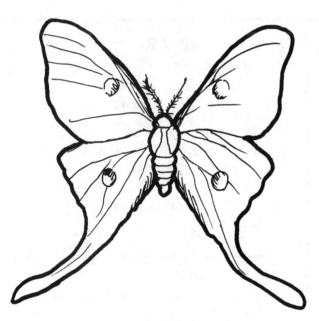

Luna Moth

Gypsy Moth

Achemon Sphinx Moth

Twin-spotted Sphinx Moth

Acrea Moth

Polyphemus Moth

Butterflies and Moths *(cont.)*

Moth Identification *(cont.)*

Luna Moth
Wingspan: 3.25 to 5" (8 to 12.5 cm)
Range: Includes eastern United States
This moth is active at night. Because of its size, shape, and color, it is one of the most spectacular of all moths.

Gypsy Moth
Wingspan: male—1.25 to 1.5" (2 to 3.75 cm), female—2.25 to 2.75" (5.5 to 7 cm)
Range: Includes eastern United States
This moth is active at night. It is mostly known as a pest of fruit and deciduous trees. It lives for only a few days or weeks.

Achemon Sphinx Moth
Wingspan: 3 to 4" (7.5 to 10 cm)
Range: Includes the entire United States
Adult sphinx moths are powerful flyers. They feed on nectar.

Twin-spotted Sphinx Moth
Wingspan: 2 to 3.25" (5 to 8 cm)
Range: Includes all of the United States
This moth is active at night. It lives on many different trees. The sphinx moth can be found from April to October.

Acrea Moth
Wingspan: 1.5 to 2.5" (3.75 to 6.25 cm)
Range: Includes the entire United States
This moth is easy to identify because of its spotted abdomen. A female acrea moth has white hindwings.

Polyphemus Moth
Wingspan: 4 to 6" (10 to 15 cm)
Range: Includes the entire United States
This moth is active at night. It only lives a few days because it does not feed.

Butterflies and Moths Specimen Box

Questions

- Do caterpillars change to butterflies?
- Do larvae change to moths?

Setting the Stage

- Show students a variety of moth and butterfly pictures.
- Ask students if anyone can identify which ones are butterflies and which ones are moths.
- Tell students butterflies and moths look alike. Quite often it is very difficult to tell them apart. Butterflies have bodies which are thin and smooth. Moths have bodies that are fat and furry. It is not easy to observe the bodies of these insects when they are flying in the air. There is another way to distinguish between the two. Look at the antennae. Moths have feathery antennae, and butterflies have club-shaped antennae.
- Ask students to observe the antennae of butterflies in their Butterfly Books. Point out the smooth edges and the bulb-shaped tips. Continue your lesson with the following activity.

Materials Needed for Each Group

- live caterpillars
- pictures of caterpillars and moths—*Butterfly and Caterpillar* by Barrie Watts (Silver Burdett Company, 1985) is a beautiful collection of color photographs.
- informational books on butterflies and moths
- live moths
- mounted butterflies and moths (optional)
- shoe box
- live plant or leaves
- plastic wrap
- transparent tape
- craft sticks
- data-capture sheet (page 64), one per student

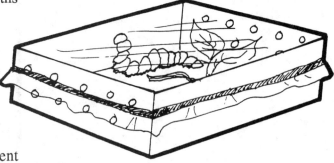

Note to the teacher: Students will construct a home for caterpillars.

Procedure *(Students Instructions)*

1. Cut an opening in the cover of the shoe box. Allow for a 1" (2.5 cm) border around the edge.
2. Cover the opening with plastic wrap. Make sure the edges are securely fastened so that the caterpillar cannot escape.
3. Punch air holes in the end of the shoe box. Do not make them too large.
4. Place the plant or leaves in the box.
5. Go outdoors and look for caterpillars for your specimen boxes. Do not touch the caterpillars with your hands. Use a twig or a craft stick.
6. Change the leaves or remember to water the plant daily. Caterpillars need fresh food.
7. On your data-capture sheet, enter observations and changes for the next several days.

Butterflies and Moths Specimen Box

(cont.)

Extensions

- Contact a lepidopterist. Ask him/her to speak with your class. Most collectors are usually honored to speak with children to acquaint them with their hobby or profession.

- Have students do research on the butterfly and moth in history. Many cultures such as the Japanese use pictures of butterflies.

- Have students make butterfly mobiles. They can use pictures in magazines. Mount them and hang them in unusual ways. Encourage students to be creative.

- Have students make butterfly kites. Cut the butterfly shape out of butcher paper or old sheets. Attach string and a tail.

- Plan nature walks. Have students take along their butterfly and moth booklets for identification purposes.

Closure

- Have students add their completed data-capture sheets in their animal journals.

- Have students create posters and make buttons to alert the school population to the plight of the animals of the rain forest.

- Encourage your students to write to the governments where rain forests are located and ask them to stop the destruction of the rain forests. Remind them that the habitat of the butterfly is being destroyed.

Butterflies and Moths Specimen Box

(cont.)

Watch your caterpillar daily, record observations, and draw pictures of the various stages it goes through.

BUTTERFLY

Caterpillar	Pupa	Butterfly	Eggs

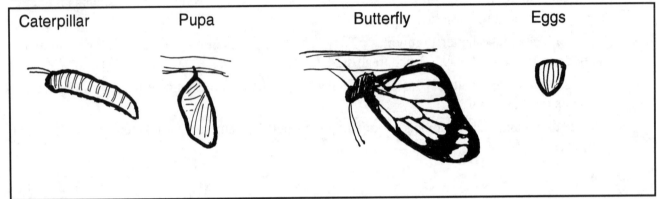

MOTH

Caterpillar	Pupa	Moth	Eggs

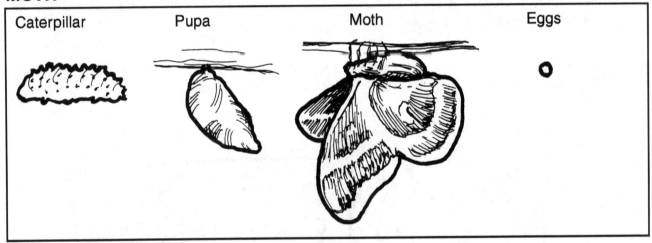

NOTES AND OBSERVATIONS:

_____

_____

_____

_____

_____

_____

_____

64

Bugs That Light the Night

Question

How does a firefly glow?

Setting the Stage

- Tell students that in the rain forest there are some insects that glow in the dark.
- Ask students if they have ever gone out on a spring or summer evening to collect tiny luminescent insects. These tiny creatures fascinate the young and old alike, but just how do these creatures glow?
- Have student do the "Bugs That Light the Night" demonstration to find out.

Materials Needed for Entire Class

- 1/4 tsp (1.25 g) Cypridina, (Cypridina are the remains of tiny marine animals which can be obtained from a biology supply house.)
- saliva or water

Procedure *(Student Instructions)*

1. In a room or a closet that is completely dark, place a pinch of Cypridina in the palm of one hand.
2. Using your other hand, add a few drops of water or saliva.
3. With your fingers mix these materials together, using a slight pressure and rotating motion to moisten completely the dried remains of the material.
4. As the material moistens, the Cypridina will glow a bright blue.

Extensions

- Take a class night trip to locate and collect other light-producing organisms. Luminescent mushrooms and other fungi, (as well as photobacteria) may be found on rotting logs in some moist areas.
- Marine dinoflagellates and other light-emitting invertebrates may be collected on the beach or in shallow salt water.
- If you do not live near a source of water, some fish in fish markets have some luminescent bacteria on their scales and mucous membranes.

Closure

In their animal journals, have students record and respond to the following questions:

How does a firefly glow? What other organisms might produce their own light? Can you find and describe some references in literature to the glowworm or firefly.

Just the Facts

The desert biome is found in almost one-fifth of all the world's land areas. Most people think of camels and sand when they think of the desert. The desert is a biome that has little or no rain all year long. In fact the desert has less than 10" (25 cm) of rainfall annually, and in most cases because of the high temperatures found there, the rain evaporates before it ever touches the ground. During the day the temperature in the desert can reach 131°F (55°C). At night the desert floor radiates the heat back into the atmosphere so that it becomes extremely cold. At night the temperatures in the desert can be as low as 32°F (0°C) or freezing. The majority of deserts were formed by the wind. The movement of sand and soil causes the desert floor to change constantly.

In spite of the extreme conditions found in the desert, animal life is abundant and varied. During the day you will not find many animals in the desert heat. There might be an occasional tortoise or snake feeding on a desert plant or a scorpion ready to sting an insect. Another animal that just might happen along is the roadrunner. Most desert animals, however, stay out of the heat during the day and come out of hiding in the cooler evenings. The skunk, coyote, and jack rabbit often find shade under a desert plant to wait out the heat of the day.

Smaller animals such as the gerbil and kangaroo rat burrow holes in the ground to protect themselves from the heat and sun. Many of the desert animals are suited to the hot climate. Their bodies are very efficient, and they can go for long periods without water.

Sometimes the animals' skin helps them survive in the desert. Most reptiles have scales that cover their bodies. Oxygen cannot pass through the scales, so reptiles do not breathe through their skin. It is also the scales that act as a moisture barrier to prevent the reptile from losing water. So they do not need water as frequently as other animals. Animals with smooth skin would quickly die in the heat of the desert because they would lose moisture through their skin.

Some animals have found very unusual ways to live and reproduce in the desert. For example, the desert toad (the spade-foot) lays its eggs in a mud puddle after a heavy rain. The eggs then hatch and develop from tadpoles to adult toads in about ten days. The whole process takes place before the mud puddle has a chance to dry up.

Mapping the Desert

Locate the world's major desert biomes and color them brown. Label the areas in which they are found.

3000 Km
3000 Mi.
Scale at the Equator.

Creating a Desert Habitat

Question

How can we keep small desert animals in the classroom?

Setting the Stage

- Tell students we can keep small animals in the classroom if we provide the correct environment for the animal. All animals need a comfortable place to live. Find a warm, sunny, protected place in the class. Be sure to choose a place that will be a permanent location. Even slight changes to an animal can upset it.
- Resource people and pet stores are a good place to find information about the type of desert animal you wish to keep. They can make suggestions on adaptability and handling. As with any type of classroom animal, be sure to caution students that some are likely to die. This is part of the natural life cycle, and we need to be aware of students' sensitivities.
- Insects and spiders have a short life span, and they eat other insects. Some students might react negatively to them.
- Good choices for the desert environment are the horned lizard, common lizard, tortoise, and small snakes. One or all can live together in a desert terrarium.

Materials Needed for Entire Class

- aquarium or terrarium, at least 20 gal (80 L)
- clean gravel
- clean sand
- plants
- assorted rocks and small pieces of wood
- shallow pan for water
- thermometer (optional)
- heat light or bottom heater (Your pet store can make suggestions.)

Procedure *(Student Instructions)*

1. Clean your terrarium.
2. Place about 1" (2.5 cm) of gravel on the bottom.
3. Add about 3" (7.5 cm) of sand on top of the gravel.
4. Arrange the rocks and twigs throughout the terrarium to provide shade and privacy.
5. Place the shallow pan with a tiny bit of water on one side of the terrarium.
6. Set heat source in place. You might want to purchase a thermometer to monitor the ideal temperature for your animal.
7. Place the animal in the terrarium. Feed it the foods recommended by the pet store.

Creating a Desert Habitat *(cont.)*

Extensions

- Have students keep notes of the animals' behavior. As they go about their daily activities, some animals might exhibit unusual behaviors that are noteworthy.
- Assign different groups of students the responsibility of caring for the pet.
- Have students keep a growth chart. Some lizards grow rapidly.
- Have students draw pictures of the animals' growth and changes.

Closure

- Have students research a variety of desert animals. Some students may have unfounded fears of snakes and spiders. Have them write their feelings in their animal journals and explain why they feel that way. How do they feel after studying about their animals?
- Arrange for a visit to a local wildlife preserve or to a museum to see these desert animals in their natural habitat.
- Have students construct desert mobiles or desert pictures. Dioramas are a good authentic assessment tool for the special needs student.

Wow! It's Hot Out Here

Question

Since the heat in the desert reaches temperatures of 131° F (54.5° C), how does a desert animal survive?

Setting the Stage

- Remind students of the temperature extremes in the desert.
- Ask students to provide answers to the survival question above. Accept any reasonable solution. Encourage them to think beyond just seeking shade. Many animals do other things to survive.

Materials Needed for Each Group

- two ice cubes of about the same size
- two small plastic bags with ties for each
- clay flower pot
- soil to fill the flower pot
- data-capture sheet (page 71), one per student

Procedure *(Student Instructions)*

1. Fill the flower pot to the rim with soil.
2. Put an ice cube into each one of the plastic bags and tie tightly with the twist ties.
3. Quickly bury one of the bags under the soil in the flower pot.
4. Place the other bag on top of the soil.
5. Put the flower pot in the sun or under a light. The light of an overhead projector is a good heat source for many experiences. (The air directly above the light is heated.)
6. Check on the melting of the ice cube above the surface of the soil. As soon as it has melted, dig the other bag out of the soil.
7. Complete your data-capture sheet.

Extensions

- Have students research which animals spend the day underground seeking protection from the heat of the day. How might this behavior affect their life styles? How would their daily life differ from that of other animals in different biomes?
- Have students make a graph of the amount of time each animal spends underground during the summer, winter, etc.

Closure

In their animal journals, have students write an acrostic about a desert animal and draw pictures of their animals and special behaviors they exhibit to protect themselves in the desert.

Sample Acrostic:

T oads
O nly lay their eggs in puddles
A t rainy times
D uring the spring or summer months.
S urvival of the species is necessary.

Wow! It's Hot Out Here *(cont.)*

Fill in the information needed.

What is the problem you are trying to solve? _____

_____

_____

What are the materials you used? _____

_____

_____

Draw a diagram or picture of how your experience looked.

AT THE BEGINNING   **AT THE COMPLETION**

Record your observations here. _____

_____

_____

_____

What did this experience teach you about desert animals? _____

_____

_____

Frogs and Tadpoles

Question

What is *metamorphosis* and what stages does a tadpole go through to become a frog?

Setting the Stage

- Tell students it is magical and mysterious. It is *metamorphosis!* Metamorphosis means change, and that is what happens to a tadpole. It changes! This change, however, is not a magic trick. It is the natural process that a tadpole goes through to become an adult. Tell students tadpoles are not the only animals that go through metamorphosis. The butterfly and moth also change. Can you think of any other cases of metamorphosis? Think of the areas of science fiction or television. In the book *Dr. Jekyll and Mr. Hyde,* the main character changes from one form to another. Can you see how metamorphosis is a great theme for a horror story?
- Ask students if people are ever transformed. You might discuss how humans change as they grow. Or how a puppy grows into an adult dog.

Materials Needed for the Class

- aquarium—5-10 gallons (20-40 L)
- pond water
- mud from the pond for the bottom of the aquarium
- tadpoles of various sizes
- data-capture sheet (page 74), one per student

Note to the teacher: Students will get a chance to observe a metamorphosis first hand. If possible, have students collect tadpoles from shallow ponds or streams. Be sure to have them collect some scummy pond water to fill an aquarium. Tadpoles need pond water to live. They feed on the bacteria that is found there. Make sure you change the water weekly so that the tadpoles have a fresh supply of food.

Procedure

1. Have students collect tadpoles and pond water in jars or pails.
2. If it is impractical to have students collect tadpoles, you can obtain them from a biology supply company. (See bibliography, pages 95-96.)
3. Have students line the bottom of the aquarium with mud. Pour the water in slowly and allow it to settle. Add tadpoles.
4. Have students observe daily so that they can record the changes in growth.

Extension

Tadpoles go through stages different from those of the butterfly. Caterpillars are easy to obtain and observe. It might be fun for your students to keep caterpillars and tadpoles to observe and compare the changes in each. A data-capture sheet similar to the one from this activity can be provided for them. Students should keep notes, drawings, and observations for this activity. They should not only record on their observation sheet, but should also record their thoughts as to what is happening in each stage.

Frogs and Tadpoles *(cont.)*

Closure

In their animal journals, have students write a science fiction story in which the main character goes through the process of metamorphosis. The setting of the story can take place in one of the areas being studied in social science.

_____
(Title)

Frogs and Tadpoles *(cont.)*

OBSERVATIONS   **DRAWINGS**

Day 1   **Date:** _____

_____

_____

_____

Day 2   **Date:** _____

_____

_____

_____

Day 3   **Date:** _____

_____

_____

_____

Day 4   **Date:** _____

_____

_____

_____

Day 5   **Date:** _____

_____

_____

_____

Just the Facts

The name *biome* is not used to describe the environment found in the oceans. Biome is a term used to describe only land environments. Scientists call the oceans and the plants and animals that live there the *aquatic habitat.* The aquatic habitat is comprised of three distinct environments, and the animals that live there are different too.

The Marine Habitat

The ocean is a great body of salt water comprising all of the oceans and seas that cover nearly three-fourths of the surface of the earth. The average depth of the ocean is 16,500' (5,000 m) and is divided by the continents. All the bodies of water that are salty make up the marine habitat. The ocean is home to a large number of sea animals.

Where an animal lives in the marine environment depends in part on the temperature of the water and the shape of the ocean floor and the coastal lands. In the marine habitat, you can find animals ranging from those as tiny as krill to those as large as the world's largest mammal, the blue whale.

Marine animals supply much of the world's food, and many people have re-creations of the marine habitat in their homes.

In the marine habitat you can find such animals as crabs, lobsters, and shrimp. Fish come in all shapes and sizes. The most colorful and unusual ones live in tropical waters. Coral reefs also live in the warm waters of the tropics. In the cooler waters off the New England coast, mussels, crabs, and scallops live.

The Freshwater Habitat

The second type of aquatic environment is the freshwater habitat. Ponds, lakes, and streams are freshwater habitats. Animals that live in the freshwater habitat are different from those in the marine habitat. Frogs, fish, turtles, and snakes are found living by ponds and streams.

The Estuary Habitat

The third type of aquatic environment is the estuary habitat. An estuary is the place where freshwater flows into the sea or an ocean. The estuary is home to many animals. These animals provide food for a great many people. Look through books and magazines to see if you can locate the different types of animal life found in an estuary.

How Do Animals of the Aquatic Live?

Mapping the Aquatic Environment

Locate the world's major aquatic environments on the world map and color them blue. Label the areas in which they are found.

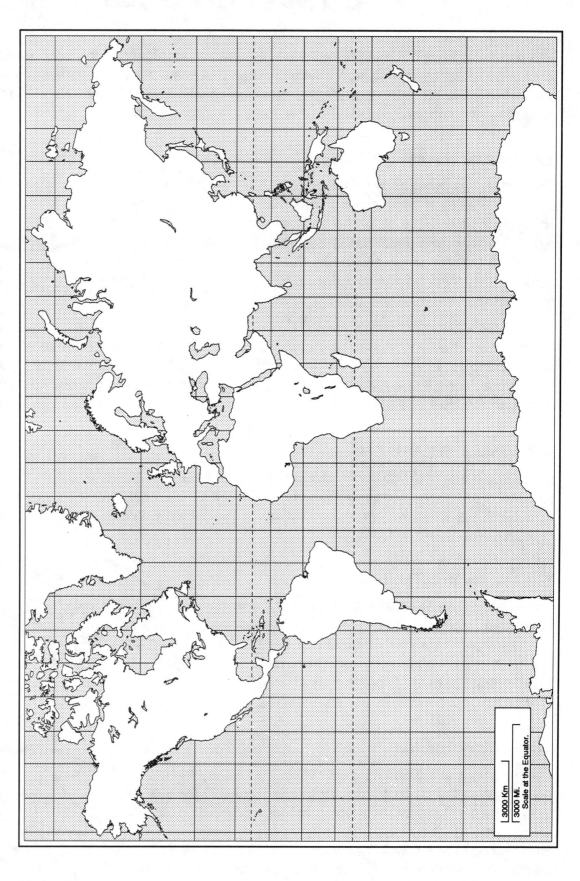

3000 Km
3000 Mi.
Scale at the Equator.

76

Fish Prints

Question

Can fish leave their prints when they do not have feet?

Setting the Stage

Pose the above question to your class. It is sort of a trick question since the prints in this activity have nothing at all to do with feet. This is a fun art activity that allows students to compare the various external structures of different types of fish.

Materials Needed for Each Group

- 4-5 different types of uncooked whole fish (A local market can assist you in purchasing these. Be sure to refrigerate.)
- black paint
- 4-5 small ink rollers or wide brushes
- white construction paper, one per student
- newspaper to cover tables
- roller trays for paint

Procedure *(Student Instructions)*

1. Prepare each paint station by covering with newspaper.
2. Pour a small amount of paint into the paint trays.
3. Roll the paint on the body of the fish.
4. Carefully place the paper on top of the fish and press gently to imprint the fish's painted body onto the paper.
5. Starting at one side, carefully peel the paper off and lay it down to dry.
6. When dry, the prints can be mounted on colored construction paper.
7. You may vary the colors used to paint the fish or even use multiple colors on the same fish. The purpose, however, is to be able to see the various scales, eye placement, and fins of different kinds of fish.

Extensions

- Have students display the fish prints as an underwater gallery in the classroom.
- Have students use the prints as covers for the aquatic habitats or for creative writing stories.
- Have students string their prints together to create unusual mobiles.

Closure

- In their animal journals, have students compare the anatomy of the various fish and enter the findings.
- Have students write to marine hatcheries and find out how they provide for the aquatic animals they raise.

By the Sea, the Beautiful Sea

Question

What does collecting seashells tell you about the animals that once called them home?

Setting the Stage

- Ask if any students have a seashell collection. If they do, they might want to display them in class.
- Provide students pictures of various sea animals and seashells. Discuss with students what similarities and differences can be noted about them. Do they live in freshwater or saltwater? Can freshwater animals live in saltwater? Why or why not?

Materials Needed for Each Individual

- a seashell collection
- numerous photos of shells and sea animals
- data-capture sheet (page 79)

Procedure

1. Display the pictures and seashells in a prominent place, visible for all the students.
2. Provide each student with a data-capture sheet.
3. Allow time for them to complete their data-capture sheets. It is a good idea to allow them to talk to one another during this activity so they can compare thoughts.

Extensions

- Conduct a field trip to a tidepool to study first-hand the animals that live there.
- Invite a marine biologist in to speak to your class.
- Have students read stories about the sea. A few are suggested in the bibliography.
- Set up an aquarium in your classroom. Fish are easy to care for, and students will love watching them. Speak to a pet store for suggestions on the hardiest fish to keep.

Closure

In their animal journals, have students write stories, acrostics, or poems about the sea.

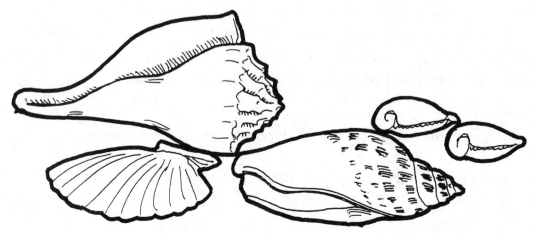

By the Sea, the Beautiful Sea *(cont.)*

Can you name the sea life in the tidepool?

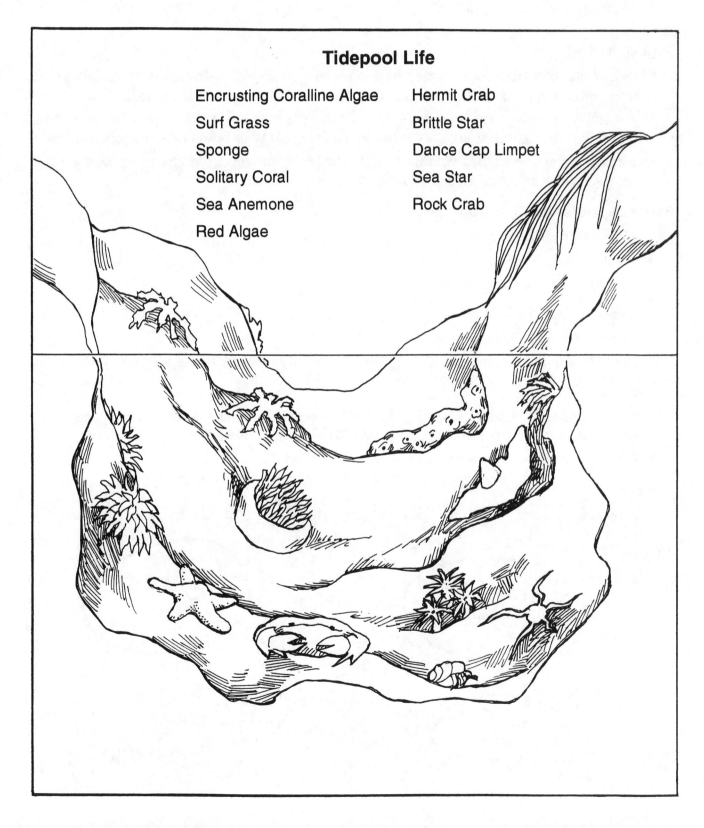

Tidepool Life

Encrusting Coralline Algae	Hermit Crab
Surf Grass	Brittle Star
Sponge	Dance Cap Limpet
Solitary Coral	Sea Star
Sea Anemone	Rock Crab
Red Algae	

Who Am I, and Why Am I Unique?

Question

Who am I, and why am I unique?

Setting the Stage

- In studying the aquatic environment, students become aware of the vast variety of animal life and the adaptations of each. Ask students to compile a list of their favorite sea animals.
- Ask students the following questions about their favorite animals. Why is it your favorite? Does it have any unusual characteristics? What does it feed on? What level of the sea does it live in?
- Play a guessing game called Who am I? Begin by giving students a few characteristics about a sea animal and letting them guess what the name of the animal is.

Materials Needed for Each Individual

- colored markers or crayons
- resource books for research
- data-capture sheets (pages 81-82)

Procedure

1. Conduct discussion suggested in Setting the Stage.
2. Have students research sea animals and complete their data-capture sheets.

Extensions

- Have the students play the guessing game using other animals from different environments. It is a quick assessment tool, and you can ascertain if further study is necessary.
- Have students draw underwater pictures with watercolors.

Closure

Have students make up their own guessing games, share them, and play them when they have extra time.

Who Am I, and Why Am I Unique?

(cont.)

Using books and resource materials, fill in the following sections about each sea animal listed.

I am the nautilus.

I live_____

_____

I move by_____

_____

Draw a picture of me and my shell.

I am a sea star.

I live_____

_____

I move by_____

_____

Draw a picture of my top and bottom side.

Who Am I, and Why Am I Unique? *(cont.)*

Using books and resource material, fill in the following sections about each sea animal listed.

We are bivalves.

Who are we? _____

_____

_____

Draw some of us and name us.

I am a gastropod.

Do you know what my common name is? _____

Do you know where I live? _____

_____

Draw a picture of me.

I am a tube worm.

Where do I live? _____

What do I look like? _____

Draw a picture of me.

Science Safety

Discuss the necessity for science safety rules. Reinforce the rules on this page or adapt them to meet the needs of your classroom. You may wish to reproduce the rules for each student or post them in the classroom.

1. Begin science activities only after all directions have been given.

2. Never put anything in your mouth unless it is required by the science experience.

3. Always wear safety goggles when participating in any lab experience.

4. Dispose of waste and recyclables in proper containers.

5. Follow classroom rules of behavior while participating in science experiences.

6. Review your basic class safety rules every time you conduct a science experience.

You can still have fun and be safe at the same time!

Animal Journal

Animal journals are an effective way to integrate science and language arts. Students are to record their observations, thoughts, and questions about past science experiences in a journal to be kept in the science area. The observations may be recorded in sentences or sketches which keep track of changes both in the science item or in the thoughts and discussions of the students.

Animal Journal entries can be completed as a team effort or an individual activity. Be sure to model the making and recording of observations several times when introducing the journals to the science area.

Use the student recordings in the animal journal as a focus for class science discussions. You should lead these discussions and guide students with probing questions, but it is usually not necessary for you to give any explanation. Students come to accurate conclusions as a result of classmates' comments and your questioning. Animal journals can also become part of the students' portfolios and overall assessment program. Journals are a valuable assessment tool for parent and student conferences as well.

How To Make an Animal Journal

1. Cut two pieces of 8.5" x 11" (22 cm x 28 cm) construction paper to create a cover. Reproduce page 85 and glue it to the front cover of the journal. Allow students to draw animal pictures in the box on the cover.
2. Insert several animal journal pages. (See page 86.)
3. Staple together and cover stapled edge with book tape.

My
Animal
Journal

Name_____

Animal Journal

Illustration

This is what happened:

_____

_____

_____

_____

_____

This is what I learned:

_____

_____

_____

_____

_____

Investigation Planner

Observation

_____

_____

Question

_____

_____

Hypothesis

_____

_____

Procedure

Materials Needed:

_____

_____

_____

_____

Step-by-Step Directions: (Number each step!)

_____

_____

_____

_____

_____

_____

Animal Observation Area

Students should be given other opportunities for real-life science experiences. For example, animal specimens and terraria can provide vehicles for discovery learning if students are given time and space to observe them.

Set up an animal observation area in your classroom. As children visit this area during open work time, expect to hear stimulating conversations and questions among them. Encourage their curiosity but respect their independence!

Books with facts pertinent to the subject, item, or process being observed should be provided for students who are ready to research more sophisticated information.

Sometimes it is very stimulating to set up a science experience or add something interesting to the Animal Observation Area without a comment from you at all! If the experiment or materials in the observation area should not be disturbed, reinforce with students the need to observe without touching or picking up.

Assessment Form

The following chart can be used by the teacher to rate cooperative learning groups in a variety of settings.

Science Groups Evaluation Sheet

Room: _____ Date: _____

Activity: _____

Everyone

	Group									
	1	2	3	4	5	6	7	8	9	10
. . . gets started.										
. . . participates.										
. . . knows jobs.										
. . . solves group problems.										
. . . cooperates.										
. . . keeps noise down.										
. . . encourages others.										

Teacher comment

_____

Bragging rights for the group session: _____

_____

Assessment Form *(cont.)*

The evaluation form below provides student groups with the opportunity to evaluate the group's overall success.

Cooperative Group Evaluation

Assignment: _____

Date: _____

Scientists	**Jobs**
_____	_____
_____	_____
_____	_____
_____	_____

As a group, decide which face you should fill in and complete the remaining sentences.

1. We finished our assignment on time, and we did a good job.

2. We encouraged each other, and we cooperated with each other.

3. We did best at_____

   _____.

4. Next time we could improve at _____

   _____.

Super Zoologist Award

This is to certify that

_____
Name

made a science discovery.

Congratulations!

_____
Teacher

_____
Date

Glossary

A

Adaptation—an alteration or adjustment in structure or habits, often hereditary, by which a species or individual improves its condition in relationship to its environment.

Amoeba—a protozoan of the genus Amoeba or related genera, occurring in water and soil and as a parasite in other animals. An amoeba has no definite form and consists essentially of a mass of protoplasm containing one nucleus or more surrounded by a delicate, flexible outer membrane.

Antennae—one of the paired, flexible, segmented sensory appendages on the head of an insect, a myriapod, or a crustacean, functioning primarily as an organ of touch.

Arid—lacking moisture, specifically having insufficient rainfall to support trees or woody plants.

B

Bacteria—any of the unicellular, prokaryotic microorganisms of the class Schizomycetes, which vary in terms of morphology, oxygen and nutritional requirements, and motility, and may be free-living, saprophytic, or pathogenic, the latter causing disease in plants or animals.

Biology—1. the science of life and of living organisms, including their structure, function, growth, origin, evolution, and distribution. It includes botany and zoology and all their subdivisions. 2. the life processes or characteristic phenomena of a group or category of living organisms.

Biome—a major regional or global biotic community, such as a grassland or desert, characterized chiefly by the dominant forms of plant life and prevailing climate.

Bivalve—a mollusk, such as an oyster or a clam, that has a shell consisting of two hinged valves.

Burrow—a hole or tunnel dug in the ground by a small animal, such as a rabbit or a mole, for habitation or refuge.

C

Camouflage—a disguise or false appearance in order to conceal; protective coloration.

Cell—the smallest structural unit of an organism that is capable of independent functioning, consisting of one or more nuclei, cytoplasm, and various organelles, all surrounded by a semipermeable cell membrane.

Chrysalis—a pupa, especially of a moth or butterfly, enclosed in a firm case or cocoon.

Class—a taxonomic category ranking below a phylum or division and above an order.

Climate—the meteorological conditions, including temperature, precipitation, and wind, that characteristically prevail in a particular region.

Cocoon—a protective case of silk or similar fibrous material spun by the larvae of moths and other insects that serves as a covering for their pupal stage.

Community—a group of plants and animals living and interacting with one another in a specific region under relatively similar environmental conditions.

Conclusion—the outcome of an investigation.

Conifer—any of various mostly needle-leaved or scale-leaved, chiefly evergreen, cone-bearing gymnospermous trees or shrubs such as pines, spruces, and firs.

Control—a standard measure of comparison in an experiment. The control always stays constant.

Glossary *(cont.)*

Desert—a barren or desolate area. A dry and often sandy region of little rainfall, extreme temperatures, and sparse vegetation.

Dormant—lying asleep or as if asleep; inactive.

Drought—a long period of abnormally low rainfall, especially one that adversely affects growing or living conditions.

Ecosystem—an ecological community together with its environment, functioning as a unit.

Estuary—1. The part of the wide lower course of a river where its current is met by the tides.
2. An arm of the sea that extends inland to meet the mouth of a river.

Experiment—a means of proving or disproving a hypothesis.

Forest—a dense growth of trees, plants, and underbrush covering a large area.

Frigid—extremely cold.

Habitat—1. The area or type of environment in which an organism or ecological community normally lives or occurs.
2. The place in which a person or thing is most likely to be found.

Hibernate—1. To pass the winter in a dormant or torpid state.
2. To be in an inactive or dormant state or period.

Humid—containing or characterized by a high amount of water or water vapor.

Hypothesis *(hi-POTH-e-sis)*—an educated guess to a question you are trying to answer.

Insect—any of numerous usually small arthropod animals of the class Insecta, having an adult stage characterized by three pairs of legs and a body segmented into head, thorax, and abdomen and usually having two pairs of wings.

Investigation—an observation of something followed by a systematic inquiry to examine what was originally observed.

Jungle— 1. Land densely overgrown with tropical vegetation.
2. A dense thicket or growth.

Kingdom—the highest taxonomic classification into which organisms are grouped, based on fundamental similarities and common ancestry.

Lemming—any various small, thickset rodents, especially of the genus Lemmus, inhabiting northern regions and known for periodic mass migrations that sometimes end in drowning.

Lepidopterist—an entomologist specializing in the study of butterflies and moths.

Luminescence—an emission of light that does not derive energy from the temperature of the emitting body. It is caused by chemical, biochemical, or crystallographic changes, the motions of subatomic particles, or radiation-induced excitation of an atomic system.

Glossary *(cont.)*

Metamorphosis—a change in the form and often habits of an animal during normal development after the embryonic stage.

Mimicry—the resemblance of one organism to another or to an object in its surroundings for concealment and protection from predators.

Observation—careful notice or examination of something.

Permafrost—permanently frozen subsoil, occurring throughout the Polar Regions and locally in perennially frigid areas.

Polar—relating to, connected with, or located near the North Pole or South Pole.

Procedure—the series of steps carried out in an experiment.

Predator—an organism that lives by preying on other organisms.

Prey—an organism that is hunted or seized for food by other organisms.

Pupa—The nonfeeding stage between the larva and adult in the metamorphosis of holometabolous insects, during which the larva typically undergoes complete transformation within a protective cocoon or hardened case.

Question—a formal way of inquiring about a particular topic.

Scientific Method—a systematic process of proving or disproving a given question, following an observation. Observation, question, hypothesis, procedure, results, conclusion, and future investigations comprise the scientific method.

Science-Process Skills—the skills needed to think critically. Process skills include observing, communicating, comparing, ordering, categorizing, relating, inferring, and applying.

Swamp—a seasonally flooded bottomland with more woody plants than a marsh and better drainage than a bog.

Taiga—a subarctic, evergreen coniferous forest of northern Eurasia located just south of the tundra and dominated by firs and spruces.

Thorax—the second or middle region of the body of an arthropod, between the head and the abdomen, in insects bearing the true legs and wings.

Tropical—hot and humid; torrid.

Tundra—a treeless area between the icecap and the tree line of Arctic regions, having a permanently frozen subsoil and supporting low-growing vegetation such as lichens, mosses, and stunted shrubs.

Variable—the changing factor of an experiment.

Zoology—the branch of biology that deals with animals and animal life, including the study of the structure, physiology, development, and classification of animals.

Bibliography

Aaseng, Nathan. *Animal Specialists.* Lerner, 1987.

Arnosky, Jim. *Crinkleroot's Book of Animal Tracking.* Bradbury, 1989.

Barton, Miles. *Animal Rights.* Watts, 1987.

Bright, Michael. *Killing for Luxury.* Watts, 1992.

Burnie, David. *Animals.* S&S Trade, 1993.

Carwardine, Mark. *Illustrated World of Wild Animals.* S&S Trade, 1990.

Cherfas, Jeremy. *Animal Defenses.* Lerner Pubns., 1991.

Cherry, Lynne. *The Great Kapok Tree, A Tale of the Amazon Rain,* Harbrace J., 1990.

Cole, Jacci. *Animal Communication: Opposing Viewpoints.* Greenhaven, 1989.

Connolly, James E. *Why the Possum's Tail Is Bare: And Other North American Indian Nature Tails.* Stemmer Hse., 1985.

Courlander, Harold. *Low-tail Switch & Other West African Nature Stories.* H Holt & Co., 1987.

Crump, Donald. *Animal Architects.* Natl Geog., 1987.

Dempewolff, Richard. *Nature Craft.* Golden Press, 1965.

DeVeto, Alfred and Gerald H. Krockover. *Creative Sciencing.* Scott Foresman, 1991.

Elson, Lawrence. *The Zoology Coloring Book.* Barnes and Noble, 1982.

Feltwell, John. *Animals & Where They Live.* Putnam, 1988.

Flegg, Jim. *Animal Communication.* Newington., 1991.

Forte, Imogene and Sandra Schurr. *Science Mind Stretchers.* Incentive Publications, 1987.

Fracklam, Margery. *Frozen Snakes and Dinosaur Bones.* Harcourt, Brace, and Javanovich, 1976.

George, Jean Craighead. *The Talking Earth.* Harper Collins, 1987.

Goodman, Billy. *Animal Homes & Societies.* Little Brown, 1991.

Griffin, Robert D. *The Biology Coloring Book.* Barnes and Noble, 1982.

Hughey, Pat. *Scavengers & Decomposers: The Cleanup Crew.* Macmillan Child Grp., 1984.

Imershein, Betsy. *Animal Doctor.* S&S Trade, 1988.

Jarrell, Randall. *Animal Family.* Pantheon, 1985.

Kalman, Bobbie. *Animal Babies.* Crabtree Pub Co., 1987.

Lester, Julius. *The Tales of Uncle Remus: The Adventures of Brer Rabbit.* Dutton/Dial, 1987.

Madden, Don. *The Wartville Wizard.* Macmillian Publishing Company, 1993.

McDonald, Joyce. *Mail-Order Kid.* Putnam, 1988.

McDonnell, Janet. *Animal Builders.* Childs World, 1989.

McDonnell, Janet. *Animal Camouflage.* Childs World, 1989.

Neisen, Thomas N. *The Marine Biology Coloring Book.* Barnes and Noble, 1982.

Oakly, Graham. *The Church Mice at Bay.* Atheneum, 1979.

Peterson, Hans. *Erik Has a Squirrel.* FS&G, 1989.

Podendorf, Illa. *Animal Homes.* Childrens, 1982.

Pope, Joyce. *Do Animals Dream?* Viking Child Bks., 1986.

Pringle, Laurance. *Animal Defenses.* HBJ, 1985.

Shipley, Rick. *The Magic of Animal Science.* Media Materials Inc., 1990.

Bibliography *(cont.)*

Silver Burdett. *Science-6.* Silver Burdett Co., 1985.

Simon, Seymour. *One Hundred & One Questions & Answers about Dangerous Animals.* Macmillan Child Grp., 1985.

Stockley, C. *Animal Behavior.* EDC., 1992.

Soto, Gary. *The Cat's Meow.* Strawberry Hill Press, 1987.

Swan, Robert. *Destination Antarctica.* Scholastic Inc., 1988.

Taylor, Barbara. *Animal Atlas.* Knopf Bks Yng Read., 1992.

Taylor, David. *Animal Attackers.* Lerner Pubns., 1990.

Tolman, Marvin N. and Morton, James O. *Life Science Activities for Grades K-8.* Parker Pub., 1986.

Venino, Suzanne. *Amazing Animal Groups.* Natl Geog., 1981.

Warner, Gertrude. *Animal Shelter Mystery.* A Whitman, 1991.

Watts, Barre. *Butterfly and Caterpillar.* Silver Burdett, 1985.

Wildlife Education. *Animal Wonders.* Wildlife Educ., 1992.

   Project Wild-Elementary Activity Guide. Western Regional Environmental Educational Council, 1986.

   Forest. Harcourt, Brace, and Jovanovich, 1990.

Spanish Titles

Byars, B. *La casa de las alas (House of Wings).* Lectorum, 1990.

Fernandez de Velasco, M. *Pabluras y Gris (Pabluras and Grey).* Lectorum, 1988.

Harvey, D. *Pescando con Pedro (Fishing with Peter).* Beautiful America, 1993.

Matrinez Gil, F. *El rio de los castores (Beaver River).* Lectorum, 1985.

Mound, L. *Los insectos (Insects).* Santillana Pub. Co., 1990.

Parker, S. *Los mamíferos (Mammals).* Santillana Pub. Co., 1989.

Technology

Bank Street College of Education. *Whales and Their Environment.* Available from Sunburst (800)321-7511. software

Focus Media. *Galactic Zoo.* Available from CDL Software Shop, (800)637-0047. software

Intentional Educations. *Life Science Topics.* Available from William K. Bradford Pub. Co., (800) 421-2009. software

January Productions. *Reptiles.* Available from CDL Software Shop, (800)637-0047. software

Lawrence Productions, Inc. *Animal Adaptations.* Available from Educational Software, (800) 421-4157. software

MECC. *Invisible Bugs.* Available from Troll, (800)526-5289. software

Orange Cherry. *Talking Jungle Safari.* Available from CDL Software Shop, (800)637-0047. software

Partridge Film & Video. *Partnerships Living in Societies.* Available from Cornet/MTI Film & Video, (800)77-8100. videodisc

Troll Associates. *Animal Facts and Adventures.* Available from Troll, (800)526-5289. software

Unicorn. *Animal Kingdom.* Available from Troll, (800)526-5289. software

WDCN, Nashville, Tennessee. *Vertebrates.* Available from AIT, (800)457-4509. video